Basketball Data Science

With Applications in R

CHAPMAN & HALL/CRC DATA SCIENCE SERIES

Reflecting the interdisciplinary nature of the field, this book series brings together researchers, practitioners, and instructors from statistics, computer science, machine learning, and analytics. The series will publish cutting-edge research, industry applications, and textbooks in data science.

The inclusion of concrete examples, applications, and methods is highly encouraged. The scope of the series includes titles in the areas of machine learning, pattern recognition, predictive analytics, business analytics, Big Data, visualization, programming, software, learning analytics, data wrangling, interactive graphics, and reproducible research.

Published Titles

Feature Engineering and Selection
A Practical Approach for Predictive Models
Max Kuhn and Kjell Johnson

Probability and Statistics for Data Science
Math + R + Data
Norman Matloff

Introduction to Data Science
Data Analysis and Prediction Algorithms with R
Rafael A. Irizarry

Cybersecurity Analytics
Rakesh M. Verma and David J. Marchette

Basketball Data Science
With Applications in R
Paola Zuccolotto and Marica Manisera

For more information about this series, please visit: https://www.crcpress.com/Chapman--HallCRC-Data-Science-Series/book-series/CHDSS

Basketball
Data Science
With Applications in R

Paola Zuccolotto
Marica Manisera

With contributions by Marco Sandri
Foreword by Ettore Messina

CRC Press
Taylor & Francis Group
Boca Raton London New York

CRC Press is an imprint of the
Taylor & Francis Group, an **informa** business
A CHAPMAN & HALL BOOK

CRC Press
Taylor & Francis Group
6000 Broken Sound Parkway NW, Suite 300
Boca Raton, FL 33487-2742

© 2020 by Taylor & Francis Group, LLC
CRC Press is an imprint of Taylor & Francis Group, an Informa business

No claim to original U.S. Government works

Printed on acid-free paper

International Standard Book Number-13: 978-1-138-60079-9 (Paperback)
978-1-138-60081-2 (Hardback)

Visit the Taylor & Francis Web site at
http://www.taylorandfrancis.com

and the CRC Press Web site at
http://www.crcpress.com

To Leonardo and Raffaele,
the best basketball players ever.

To Alberto,
(whatever the sport) MVP.

To all three of them, budding Data Scientists.

Contents

PART III Computational Insights

Foreword

When we are "at home" in San Antonio, I arrive in the office very early and the first email I read is the FaxBack, or the statistics relating to the next opponent and the comparison with the Spurs.

Invariably, I can't help smiling to myself, remembering that I am undoubtedly one of the "old school" coaches who has been given a cleanup in the last few years, learning, with a certain degree of pride, to converse with our analyst staff. I have learned not only to utilize the work of this group of five people who provide us daily with a mountain of data, but also I have devised my own set of criteria which help me to navigate my way through all the information, selecting those items which allow me to prepare for the game, studying our opponent's strengths and weaknesses. I study the statistics which give me a better understanding of their "tendencies" and what they are likely to do in the crucial moments of the game.

In recent years, the National Basketball Association (NBA) teams have strengthened their analyst staff to prepare in the best way for the draft, to carry out player exchanges, and to study the most efficient attack and defense aspects of the game. This is to the point where, in some teams, the style of game sought by the coaching staff has been predominantly influenced by numbers. It is no secret that many of the teams have started requiring their players to aim for a shot from underneath the basket or from behind the three-point line, considering the middle distance shot totally inefficient. I am convinced that this could be a guideline but not a diktat; in other words, if all those in attack go for shots from below and shots from the three-point line, the defenders will do their best not to concede these shots. As a result, they may allow more shots from the middle. At this point, having some athletes in the team who can shoot well from this distance will become absolutely a must in order to win.

Personally, I believe that the human factor is always fundamental: first of all, speaking of simple things, I am interested in knowing, as well as the point averages, rebounds, turnovers, assists, how much of all this is done against the best teams. Too many players who average ten points per game for the season score most of those points against average and poor teams, whereas they are less successful against the top teams. Also, scoring ten points in the first two quarters of the game is not the same as scoring them in the final minutes when the outcome of the match is being decided.

Furthermore, the coach's decision as to whom to put on the field is based above all on feelings about the character of the individual players and their personal chemistry with the teammates, rather than exclusively on their technical skills, and about their ability to cope with pressure. Numbers can, and should, help us. However, communication between coaches and analysts can only be the central point of efficient management of resources, which are, before anything else, human.

I am grateful to Paola Zuccolotto and Marica Manisera for sharing this "philosophical" approach in their valuable work. I think that it is the correct route for bringing these two worlds closer together and achieving the maximum pooling of knowledge.

Lastly, I hope that they will let me have a copy of their book soon so that I can pass it on to the Spurs analysts. Italians do it better!

Ettore Messina
San Antonio, February 2019

Ettore Messina is an Italian professional basketball coach, currently Head Coach of Olimpia Milano, Italy. From 2014–2019, he has been Lead Assistant Coach of Gregg Popovich with the San Antonio Spurs of the National Basketball Association (NBA) and, before that, served as an assistant to the Los Angeles Lakers head coach Mike Brown in the 2011–2012 season. He previously coached professionally in Europe for more than 20 years.

He has won four Euroleague championships as a head coach, with Virtus Bologna in 1997/98 and 2000/01 and with CSKA Moscow in 2005/06 and 2007/08.

From 1993–1997 and in 2016–2017, Ettore Messina was the head coach of the senior men's Italian national basketball team. He coached Italy at the EuroBasket 1993, 1995, 1997 and 2017, and he also led Italy to the silver medal at the EuroBasket 1997.

Ettore Messina is regarded as one of the best European basketball coaches of all time, having been named one of the 50 Greatest Euroleague Contributors in 2008. He was named the Italian League's Coach of the Year four times (1990, 1998, 2001 and 2005) and four straight times in the Russian League (2006, 2007, 2008, and 2009). Furthermore, he has been named Euroleague Coach of the Year three times (1998, 2006, and 2008). He was inducted into the Italian Basketball Hall of Fame in 2008, in the Russian VTB League in 2019, and named one of the 10 Greatest Coaches of the Euroleague in 2008.

Preface

We began analyzing sports data several years ago with some experiments on soccer. Together with a number of colleagues, we developed a statistical algorithm aimed at filtering out the random factors due to alternating fortune and redefining match results based purely on performance on the basketball court. Apart from some results of minor impact, a slight reordering emerged at the top of the final ranking of the Italian championship, with the result that, according to Data Science, the Scudetto (the prize given to the champions of Italy's top division of professional soccer) should have been awarded more deservedly to the team ranked second. Naturally enough, this event ended up in the newspapers, exacerbating the hostility which already existed amongst soccer supporters. We did not go any further into the research.

We recommenced work on sports around three years ago focusing mainly, largely by chance and more than anything for fun, on basketball. After a few months, we became aware of the existence of an unexpected variety of directions in which to focus our research, of a limitless number of questions awaiting a response, and a moderate wealth of interesting facts which were reasonably easy to access.

Captivated by the idea of pursuing this new avenue, we got in contact with colleagues all over the world who were carrying out scientific research on sports-related subjects and found that we were in good company. However, it was not enough for us. From the very beginning, it was clear to us that contact with technical sports experts would allow us to target our efforts better in the right direction and to assess the results. We therefore came up with the idea of creating a project, which we have called Big Data Analytics in Sports (BDsports, https://bdsports.unibs.it), based on the Big & Open Data Innovation Laboratory (BODaI-Lab, https://bodai.unibs.it) of the University of Brescia, Italy, and setting up within it a network of individuals concerned with sports analytics with the specific objective of putting the academic world in contact with the world of sports. BDsports was established in the summer of 2016 while the Olympics were being staged in

Rio de Janeiro and, as of today, numerous activities have been undertaken by members of the team in the four channels which characterize the project's activities: scientific research, implementation on the ground, teaching and dissemination. In its mission, BDsports is open to all sports: today it is working on soccer, volleyball, athletics, tennis, fencing, dancesport, as well as, naturally, on basketball, which we have both continued to work on exclusively. We have published both scientific articles in international journals and informative articles in specialized newspapers and magazines. We have organized workshops and sessions at scientific conferences, as well as meetings with the public and industry specialists. We have created our own network of contacts with the world of basketball which we aim to continue building on, with the objective of disseminating Data Science culture at all levels; that is, the ability to extrapolate information from the data and to interpret it, bearing in mind the potential and limits of the methods used. We have devoted special attention to young people because, tomorrow, the world of sports will be in their hands. Thus, we provide 360-degree training: from teaching to tutoring of master's theses and PhD dissertations, to the organization of internships.

At the moment, we are on the eve of another remarkable breakthrough: the International Statistical Institute (ISI) delegated to us the task of constituting a new Special Interest Group on Sports Statistics in order to further extend our network all over the world under the aegis of the most authoritative international organization in the field of Statistics operating worldwide since 1885. This mission is a great honor for us and will hugely help us to build awareness, disseminate and establish the culture of Sports Analytics.

This book is an important stage in our journey. We have assembled our experiences in the analysis of basketball data and have tried to talk about them in a language which is technical but also, where possible, simple in order to communicate with a wide range of interlocutors.

Simultaneously with the writing of the book, together with our colleague Marco Sandri, we developed the R package BasketballAnalyzeR, which has grown alongside the book itself, following the needs of the case studies that we wanted to describe. A big part of the presented examples are relative to one single team, the Golden State Warriors (however, codes can be easily customized to be used with other teams' data). The idea behind this choice is that the reader can imagine to be interested in a given team, perform a huge set of analyses about it and, in the end, build a deep awareness from a lot of different perspectives by summing up all the results in mind. Thanks to `BasketballAnalyzeR` everyone is given the ability to reproduce the analyses set out throughout the book and the possibility of replicating them on other data. The functions of the package have been structured in such a way as to be able to be used at different levels, from elementary to advanced, with customizations and outputs on which to base any subsequent analyses. Each chapter contains both simple analyses and ideas for more complex in-depth studies, contained in a final section called FOCUS, which gives a glimpse of topics relating to scientific research.

A distinctive feature of this book is that it is specifically and exclusively concerned with basketball, although many of the arguments could be extended by analogy to other sports.

At whom is this book aimed?

As set out above, the whole book has been structured so as to be able to achieve the ambitious objective of addressing a range of different audiences. As fundamental statistical concepts are often cited, with the assumption that they are already known to the reader, a necessary requisite for tackling the issues in this book is good familiarity with the principal notions of Statistics at the level they are usually taught in basic university courses. A good knowledge of the R language is also essential to be able to understand the codes provided and to implement them to reproduce analyses on their own data. However, a possible lack of this knowledge does not affect the reading of the book. In detail, this book is aimed at:

- *Students of advanced university courses, PhDs and master's.* For students in Data Science courses, it could constitute a collection of fairly simple analyses, aimed at introducing them to, and guiding them through, the world of basketball analytics, whilst students in courses aimed at specific sports training could find a useful group of tools which can be applied immediately on the field.

- *Technicians, sports coaches and analysts.* The availability of R codes means that the suggested analyses can be reproduced on their own data, providing support tools to decisions in terms of training and definition of game strategies or creative ideas for match and championship commentary.

- *Data Scientists concerned with advanced development of sports analytics.* For Data Scientists, the book could constitute an introduction to applications in the field of basketball, an initial knowledge base on which to stimulate suggestions and ideas for further research. The functions of the R package BasketballAnalyzeR provide outputs of objects and results which could be used for more in-depth analysis.

How to read this book

Since the range of different audiences is wide, the book is structured into three levels of complexity from the point of view of the statistical techniques. Nevertheless, compatibly with each level, the writing always remains as simple as possible in order to allow everybody to cope with the essence of the key issues.

- *Part I: Introductory level.* In this part, only basic statistical methods and indexes are addressed, also with the detailed explanation of some fundamental statistical concepts.

- *Part II: Intermediate level.* In this part, more complex methods are addressed. For each method, an essential explanation is given in simple terms, but the most technical details about the statistical issues are deliberately skipped. For each method, references for self-study are given.

- *FOCUS sections: Expert level.* The FOCUS sections in the end of some chapters report on results contained in scientific papers, which can be of high complexity. Simple-term explanations are still given, but a deep understanding is left to highly skilled readers who can refer to the cited scientific papers.

Exercises can be made by reproducing all the presented analyses on different datasets that can be bought from data providers, retrieved from the web or even created ad hoc. Some datasets are available at

https://bdsports.unibs.it/basketballanalyzer/

in the section "Data for exercises". The last chapter gives some suggestions on how to start from different data sources and adapt them to the format required by BasketballAnalyzeR.

Acknowledgments

Thanks go to Raffaele Imbrogno, statistician, match analyst, trainer and technician of the highest level, for the torrent of ideas with which he has stimulated our creativity daily, and for the intensive work he does to bring the culture of Data Science into the world of basketball.

An enormous thank you to coach Ettore Messina for having written the Foreword to the book in words which convey all the emotion felt when great basketball is played on court and the strong passion and enthusiasm of the NBA world. We thank him also for the confidence that he has placed in this project: if he believes in it, then we also believe in it.

Thanks to Marco Sandri, who has been a great friend and companion in scientific research for 25 years. We have had all sorts of adventures in the application of Data Science, but this is something we could never have dreamt of. It is thanks to his skill, precision and dedication that the R package `BasketballAnalyzeR` has become a reality.

Thanks to our colleagues/friends: Christophe Ley, Andreas Groll, Leonardo Egidi, Maurizio Carpita, Nicola Torelli, Francesco Lisi, Luigi Salmaso, Livio Corain, Stefania Mignani, Alessandro Lubisco, Ron Kenett, Gerhard Tutz, Maria Iannario, Rosaria Simone, Gunther Schauberger, Hans Van Eetvelde, Manlio Migliorati, Ioannis Ntzoufras, Dimitris Karlis, and all the members of BDsports who share with us the enthusiasm of this venture.

Thanks to Fabrizio Ruggeri for actively involving us in the International Statistical Institute (ISI) and to ISI for entrusting us with the new challenge of constituting a new Special Interest Group on Sports Statistics. Thanks to all those who have joined the group (in addition to the already mentioned BDsports members, we'd like to thank Jim Albert, James Cochran, Philip Maymin and Tim Swartz) or promised us their support, in particular Luke Bornn, Scott Evans and Mark Glickman. Scientific research also progresses thanks to these important networks of scholars.

Thanks to Paolo Raineri, CEO of MYagonism, because he never stopped believing that Data Science could be brought into basketball, even when nothing seemed to go the right way. Thanks for the passion which he puts into it and for the exchange of ideas (and information) which have helped us to grow.

If the results of the statistical analyses were a special cake made by a skilled *patissier*, the data would be the ingredients. Thanks to Guido Corti, President of AIASP, for the constant commitment to the theme of data quality in basketball, which is a bit like guaranteeing premium ingredients to the *patissier*. Thanks to Serhat Ugur, founder of BigDataBall, for offering high-quality datasets for basketball analytics and making them available to us for this book and the R package `BasketballAnalyzeR`. We also thank him for his enthusiastic support of our BDsports project.

Thanks to coach Marco Crespi for our exchange of ideas some years ago. It was then, for the first time, that with his groundbreaking approach to basketball, he convinced us that a dialogue could and should be established between the world of scientific research and the basketball played on court.

Thanks to Roberta Tengattini, Davide Verzeletti, Mael Le Carre, Gianluca Piromallo and all the other students who we have had the pleasure of tutoring in their bachelor's and master's theses and university internships, in the master's courses devoted to Sports, in the PhD courses on Sports Analytics, in the specialized courses organized for coaches and technicians, because each time you teach someone something, you learn it a little better yourself.

Thanks to everyone, including those with whom we have had just a few chats on the subject of Statistics in basketball, because the wealth of ideas on which this book is based is like a puzzle in which even the smallest pieces are essential.

Sincere thanks to David Grubbs, for inviting us to write this book. We immediately got caught up in all the excitement he conveyed to us about this enterprise. Many thanks also for his (and the whole CRC Press team's) highly professional assistance. In addition, we wish to acknowledge the great efforts of the reviewers, Jack Davis, Patrick Mair, Jacob Mortensen, Sergio Olmos, Jason A. Osborne, Virgilio G. Rubio and Brian Skinner, who thoroughly read the book and granted us a multitude of extremely valuable suggestions that have greatly improved our work.

The workplace merits a special mention because more work is done and a higher standard of work is achieved when there is a friendly and calm environment. Therefore, thanks go to all our colleagues, our "work family". Warm thanks go to Aride Mazzali and Domenico Piccolo for everything they taught us and to Eugenio Brentari and Maurizio Carpita for always being those we can rely on. And an affectionate thought is given to Livia Dancelli who is no longer there but in truth is always there.

Last, but not least, heartfelt thanks go to Nicola and Giulio, for so many reasons that the book is not big enough to list them all, and to our families for having always given us their support. And now, let our mothers' hearts thank our sons, Leonardo and Raffaele, because every morning at breakfast they update Paola on the most important events which have happened overnight in the NBA championship. Many of the examples shown in the text come from some idea initiated by them, which are always a great help when it comes to interpreting the results of some analysis. Alberto is still too young for all of this, but over these years he has heroically endured Marica's time being dedicated not completely to him but to producing the drawings on the computer which are not even very attractive and certainly not nearly as good as Alberto's masterpieces.

Paola Zuccolotto and Marica Manisera

Authors

Paola Zuccolotto and **Marica Manisera** are, respectively, Full and Associate Professors of Statistics at the University of Brescia, Italy.

Paola Zuccolotto is the scientific director of the Big & Open Data Innovation Laboratory (BODaI-Lab), where she coordinates, together with Marica Manisera, the international project Big Data Analytics in Sports (BDsports).

They carry out scientific research activity in the field of Statistical Science, both with a methodological and applied approach. They authored/co-authored several scientific articles in international journals and books and participated in many national and international conferences, also as organizers of specialized sessions, often on the topic of Sports Analytics. They regularly act as scientific reviewers for the world's most prestigious journals in the field of Statistics.

Paola Zuccolotto is a member of the Editorial Advisory Board of the *Journal of Sports Sciences*, while Marica Manisera is Associate Editor of the *Journal of Sports Analytics*; both of them are guest co-editors of special issues of international journals on Statistics in Sports.

The International Statistical Institute (ISI) delegated to them the task of revitalizing its Special Interest Group (SIG) on Sports Statistics. Marica Manisera is the Chair of the renewed ISI SIG on Sports.

Both of them teach undergraduate and graduate courses in the field of Statistics and are responsible for the scientific area dedicated to Sports Analytics in the PhD program, "Analytics for Economics and Management" in the University of Brescia. They also teach courses and seminars on Sports Analytics in University master's Sports Engineering and specialized courses in training projects devoted to people operating in the sports world.

They supervise student internships, final reports and master's theses on the subject of Statistics, often with applications to sports data. They also work in collaboration with high school teachers, creating experimental educational projects to bring students closer to quantitative subjects through Sports Analytics.

I

Getting Started Analyzing
Basketball Data

a

CHAPTER 1

Introduction

A CCORDING TO Marc Gasol, statistics are killing basketball. The Spanish big man gave his opinion about statistics almost at the end of the regular season of the NBA 2016/2017, when, according to statistics, he had just become the first center to record 300 assists, 100 threes and 100 blocks in a single season.

> We've got 43 wins. If we win (tonight), we'll have 44. That's the only number you guys (media) should care about. Stats are great, but wins and losses matter. Stats are killing the game of basketball. Basketball is a subjective game. A lot of things happen that you cannot measure in stats. Different things matter. To me, the most important things in basketball are not measured by stats.
>
> Marc Gasol
> #33 Memphis Grizzlies 2016/2017

Let's start our book on basketball Data Science by providing a disconcerting revelation: according to us, Marc Gasol is right. What we are talking about is Statistics and Data Science in basketball and how to make it a useful tool but not, as Marc Gasol fears, as a way to reduce the game to numbers that are not truly able to describe it. So, let's start from the wrong way of understanding Sports Analytics. First, the vast majority of people believe that Statistics in basketball can be reduced to counting the number of shots, baskets, points, assists, turnovers, ... And, in a way, this simplification makes sense: every day the specialized media report news about these so-called statistics, detected in the NBA games, and fans are delighted to bet on whom will be the first player who will exceed this or that record. But these statistics (and we deliberately continue to write the term with a lowercase initial letter, to distinguish it from Statistics, which is the science we are dealing with) don't say much, and Gasol is right. For a Statistician, these are simply data that, collected in large quantities and appropriately re-elaborated, can be transformed into useful information to support technical experts in their decisions. Surely, to evaluate a performance only on the basis of these values is not only very reductive but even, in some cases, misleading. In addition, we think that the concept that Marc Gasol tried to convey is one of the most crucial: the media, with their often sensationalist claims, spread the statistics as if they were the thermometer of the players' skills and success. In this way, players will sooner or later change their way of playing with the goal of keeping high stats, with all due respect to teamwork. And the only real goal is to win. Actually, the opinion of the Spanish player is not limited to this topic, already fundamental in itself, but adds another aspect that is even more subtle: are the numbers really able to describe the game? How could we measure aspects such as the way a point guard is able to control the pace of the game and his decision-making skills, the influence of a leader on the team's self-confidence, the cohesion of the players, the extent to which defense is firm and tough, etc.? Certainly this is not counting the number of assists, points, and steals. You do not need to be an expert to understand it. On this point, unexpectedly, Statistics (the one with a capital S) has a lot to say because there is a well-developed research line that deals with the study of latent variables, that is all those variables that are not concretely and physically measurable. The tools invoked by these methods are sophisticated techniques and algorithms. So, if Big Marc ever read this book, he would find out that he is right: the most important facts of basketball are not measured by statistics. But they could be measured by Statistics and, in general, by Data Science.

1.1 WHAT IS DATA SCIENCE?

Data Science is the discipline aimed at extracting knowledge from data in various forms, either structured or unstructured, in small or big amounts. It can be applied in a wide range of fields, from medical sciences to finance, from logistics to marketing. By its very nature, Data Science is multidisciplinary: it combines Statistics, Mathematics, Computer Science and operates in the domains of multivariate data analysis, data visualization, artificial intelligence, machine learning, data mining, and parallel computing. In fact, several skills and abilities are required for a Data Scientist: he needs to be familiar with Computer Science, as he has to handle complex databases in different formats from different sources and to use or develop codes to run algorithms; Statistics and Mathematics are then necessary, to the aim of extracting knowledge from data through more or less sophisticated methods and models; furthermore, a Data Scientist greatly benefits from some expertize in the application field he is working on, in order to ask the right research questions and translate them into hypotheses to be tested with statistical methods.

1.1.1 Knowledge representation

There are several different ways for representing the patterns that can be discovered through Data Science, and each one dictates the kind of technique that can be used to infer the selected output structure from data (Witten et al., 2016). A broad spectrum of categorizations are proposed in the scientific literature.

For example, according to Witten et al. (2016), the structural patterns can be expressed in the form of tables, linear models, trees, rules (classification rules, association rules, rules with exceptions, more expressive rules), instance-based representation, and clusters.

In Han et al. (2011), the kind of patterns that can be discovered by data mining functionalities are associations and correlations, classification and regression for predictive analysis, cluster analysis, and outlier analysis.

Similarly, Larose and Larose (2014) states that the most common tasks that can be accomplished by data mining techniques are description, estimation, prediction, classification, clustering, and association analysis.

In a book specifically devoted to Data Science for sports analytics, Miller (2015) states that doing data science means implementing flexible, scalable, extensible systems for data preparation, analysis, visualization, and modeling.

A quite different point of view is proposed by Information Quality (InfoQ, defined by Kenett and Shmueli, 2016), a paradigm concerned with the assessment of the potential of a dataset for achieving a particular analysis goal. InfoQ proposes three general classes of goals: description (quantifying population effects by means of data summaries, data visualization techniques and other basic statistical methods), causal explanation (establishing and measuring causal relationships between inputs and outcomes) and empirical prediction (forecasting future values of an outcome based on the past and a given set of input variables). According to InfoQ, the quality of information is determined by eight dimensions, related to data quality, temporal relevance, causality, generalizability, findings operationalization and communication (Kenett and Redman, 2019). Such dimensions are founded on a full understanding of the problem and the context in which it occurs and allow the Data Scientist to provide useful results and recommendations to the decision maker and help define how such results are operationalized (that is put into practice) and communicated. Among the eight dimensions, it is worth mentioning here the generalizability of results, viewed as statistical and scientific generalizability. Statistical generalizability refers to issues related to the ability of a sample to represent an entire population and the possibility to predict the values of new observations and forecasting future values, according to a predefined goal. The statistical literature is very wide on this topic, and always new methodologies are proposed. On the other hand, a key role is played by scientific generalizability, in the sense of reproducibility, and/or repeatability, and/or replicability. For example, in basketball studies, it is very important to keep in mind that results obtained with NBA data are not always immediately generalizable to Europe (space generalizability), as well as patterns about relationships among players, or performance or game strategies can be valid only for some layups, teams, or championships (type generalizability), or in certain game situations (setting generalizability) or in some periods of time (time generalizability).

Of course, it is impossible to mention a full list of approaches in this sense and a unique categorization shared by all scientists. In this book, we will firstly consider the very general issue of discovering patterns— dealing with the broad concept of finding hidden regularities in data and including a wide range of different topics—and we will then focus on the two more specific problems of finding groups and modeling relationships, with reference to basketball applications.

1.1.2 A tool for decisions and not a substitute for human intelligence

Statistics is a subject unknown to most, often reduced to statements concerned with simple counts or percentages or, right in the luckiest cases, recalling the concept of probability. But very rarely people recognize its true meaning of "science of extracting information from data" and are aware of its enormously wide range of applications. Data Science is an even more misunderstood topic; the only difference is that these days it is such a fashionable buzzword. But buzzwords and suspicion often go hand in hand, especially for those topics hard to understand without a proper skill set. So, we often come across funny quotes about resorting to Statistics with the aim to lie, to support one's previous beliefs and to overshadow feelings and intuitions, replacing them with cold calculations.

Statistics are used much like a drunk uses a lamppost: for support, not illumination.

Vincent Edward Scully
American sportscaster

This mistrust of algorithms frequently affects the field of Sports Analytics, where technical experts may feel marginalized by obscure methods, unable to give due credit to their expertise.

So, to introduce this book, it is worth clarifying some key concepts concerning Data Science.

- **Data Science aims at extracting knowledge from data**. Starting from a research question, data that are supposed to contain the information necessary to provide an answer are collected and processed; the final phase consists in interpretation and deployment.

- **Interpretation of results is an extremely delicate phase**. William Whyte Watt stated in his book, "An American Rhetoric," that one should not trust what Statistics says without having carefully considered what it does not say. This means that the results of a statistical analysis depend on the assumptions and the available data, so we should not expect to extend their significance beyond these boundaries, and attribute to them conclusions that they are incapable to accredit. In addition, we should always remember that results are more stable, robust and reliable when the data set on which they are based is larger.

- **Data Science can deal with any field of human knowledge** (medicine, economics, finance, genetics, physics, etc.), so the Data Scientist needs to have some expertise in the application field he is working on, but of course he cannot be an expert on anything, as well as experts in the application fields cannot also be Data Scientists. It goes without saying that the Data Scientists must work side-by-side with the experts, whose task is formulating the research questions, helping to identify the data that contain the information and interpreting the obtained results. In other words, teamwork is essential.

- **Potentially, there is no question that Data Science cannot answer if it has the right data.** Sometimes these data exist and we just need to find a way to recover them. Sometimes, they do not exist, but an experimental design to collect them can be programmed: in this case, we must be patient and wait all the time it takes to generate the data. Sometimes, data do not exist or collecting them can be impossible, too difficult, expensive or time consuming. In this case, we must necessarily compromise and accordingly change the research question (as claimed by Hand, 2008, "Statisticians [...] may well have to explain that the data are inadequate to answer a particular question."). Almost always we face a combination of these situations, so if we ask ourselves whether Data Science can really answer all the questions, we should say "yes, in theory, but not in reality".

- **Data Science will never be able to describe everything of the analyzed topic.** This point is a natural corollary of the previous one: we must be aware of the presence of a broad spectrum of qualitative and intuitive overtones, not traceable throughout quantitative approaches (mostly due to the lack of proper data, not of proper methods).

- **Data Science is not a crystal ball.** It does not provide certainties (be wary of those who claim to be able to do that!) but likely scenarios and medium-long period structural indications.

- **Data Science alone does not provide decisions, but support for decisions.** Following all the previous remarks, we conclude that algorithms can never replace the human brain when it is necessary to pull the strings of all the evidence and formulate the final judgment. This is because only the human brain is able to summarize all the gathered information, bringing together the quantitative and qualitative issues in a single perception, at the same time taking into account all the imponderable and unconscious elements that cannot be formalized in a structured way.

Therefore, Data Science has no ambition to replace basketball experts, but rather aims to support them in their choices and decisions, trying to give a quantitative answer to the questions they pose. Figure 1.1 illustrates what we call the virtuous cycle of Sports Analytics, where basketball experts (stylized men with hats) and Data Scientists (stylized men with glasses) collaborate with a common aim. The basketball expert, as already mentioned, formulates the problem, by posing the right research questions, which allow the Data Scientist to understand the problem and narrow the focus. Then, basketball experts together with Data Scientists plan the research design. In this step, their cooperation is fundamental to understand which data to consider, along with their strengths and limitations. Subsequently, the analyst carries out the analysis, using his statistical competences, and then the ball goes back to both of them for a joint interpretation of the results. The information and knowledge drawn from the data thanks to Data Science is the support for the basketball experts' decisions, on whom the decision responsibility lies. They must summarize all the information available to them: the quantitative evidences provided by Data Science and the qualitative ones deriving from their intuitions which, coming from expe-

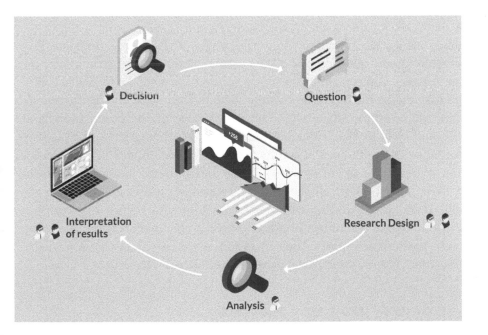

Figure 1.1 The virtuous cycle of Sports Analytics. Artwork by Gummy Industries Srl (https://gummyindustries.com/).

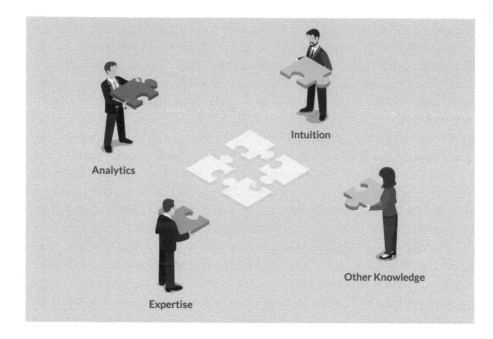

Figure 1.2 Anatomy of a decision. Artwork by Gummy Industries Srl (https://gummyindustries.com/).

rience, are no less important than the others. The experts' decisions give then life to new questions, and the cycle continues. Mark Brunkhart, the Los Angeles Galaxy Data Scientist, pointed out that Data Science cannot take credit for wins: the team is the one who wins. But Data Science gives the team an advantage. The perspective of the coach is right at 90% and it is in that remaining 10% that Mathematics and Statistics can make the difference. Therefore, Data Science should not be considered the report card of coaches, teams and players, but an instrument at their service, with all the potentialities and limits set out above, to understand strengths and weaknesses. As shown in Figure 1.2, analytics is just a piece of a complex puzzle composing the final decision, that is up to the basketball experts: their intuition, expertise and other knowledge play a key role in composing the final decision in the expert's head.

1.2   DATA SCIENCE IN BASKETBALL

Data Science applied to sports data is gaining a rapidly growing interest, as demonstrated by the really huge number of books published on this topic in the last years (Albert et al., 2017; Severini, 2014). More and more coaches, players, scouts and sport managers recognize its value as a support for decisions, and a quantitative approach to sport is often found funny and appealing also by fans and sportscasters (Winston, 2012; Alamar, 2013). A comprehensive approach to Sports Analytics includes an understanding of the whole world centered around a sport, *i.e.*, the industry, the business, and what happens on the fields and courts of play. In this book, we focus attention on the latter aspect, concentrating on the issues related to the players, the playing patterns, the games and the factors influencing performance. Topics such as managing sports markets, marketing communications, brand development, finance, also using modern Data Science approaches concerned with social network analysis, text and sentiment analysis can be found in Miller (2015).

In the strict context of player and game analysis, there are sets of basic tools (typically, indexes and charts) that are considered the gold standard and are more or less (depending on the sport and on the professional level) commonly used by experts. Often there are websites where these indexes and graphs are computed, constantly updated and made available to everybody (at least for the most famous championships, teams, athletes) or upon registration. In the case of basketball, a milestone of such analytics is Dean Oliver's pioneering book (Oliver, 2004), together with his longtime website (www.rawbw.com), Journal of Basketball Studies. Oliver's framework for the evaluation of players and teams is probably the most commonly resorted to by experts and practitioners, thanks to his research into the importance of pace and possessions, the influence of teamwork on individual statistics, the definition of offensive and defensive efficiency ratings, the development of the Four Factors of Basketball Success (field-goal shooting, offensive rebounds, turnovers and getting to the free-throw line, Kubatko et al., 2007). These analytics are complemented by a wide set of further indexes, which have become customary in the current basketball analytics jargon (the Total Basketball Proficiency Score of Kay, 1966; the Individual Efficiency at Games of Gómez Sánchez et al., 1980; the Magic Metric developed by the Mays Consulting Group, MVP IBM, just to cite a few). In this context, Schumaker et al. (2010) identifies five main topics in basketball analytics: shot zones analysis, player efficiency rating, plus/minus rating, measurement of players' contribution to winning the game, and rating clutch performances. Regarding NBA games and players, a surprisingly vast spectrum of analytics, both indexes and charts (often customizable by the user), can be found on several websites, such as, to give but few examples, NBA.com, ESPN.com, sportingcharts.com (see Kubatko et al., 2007) and cleaningtheglass.com. It would be impossible to draw here a complete list of the websites and the (more or less) educational books dealing with these issues. Nevertheless, to examine and discuss such well-established tools is outside the scope of this book, whose focus is, instead, on the use of Data Science to answer specific basketball questions. Examples of this kind of quantitative approach can be found in the scientific literature, where several statistical techniques have been applied to investigate complex problems, with a great variety of different aims, such as:

1. predicting the outcomes of a game or a tournament (West, 2008; Loeffelholz et al., 2009; Brown and Sokol, 2010; Gupta, 2015; Lopez and Matthews, 2015; Ruiz and Perez-Cruz, 2015; Yuan et al., 2015; Manner, 2016; Vračar et al., 2016),

2. determining discriminating factors between successful and unsuccessful teams (Trninić et al., 2002; Sampaio and Janeira, 2003; Ibáñez et al., 2003; De Rose, 2004; Csataljay et al., 2009; Ibáñez et al., 2009; Koh et al., 2011, 2012; García et al., 2013),

3. examining the statistical properties and patterns of scoring during the games (Gabel and Redner, 2012; Schwarz, 2012; Cervone et al., 2016),

4. analysing a player's performance and the impact on his team's chances of winning (Page et al., 2007; Cooper et al., 2009; Sampaio et al., 2010a; Piette et al., 2010; Fearnhead and Taylor, 2011; Özmen, 2012; Page et al., 2013; Erčulj and Štrumbelj, 2015; Deshpande and Jensen, 2016; Passos et al., 2016; Franks et al., 2016; Engelmann, 2017), also with reference to the so-called "hot hand" effect (Gilovich et al., 1985; Vergin, 2000; Koehler and Conley, 2003; Tversky and Gilovich, 2005; Arkes, 2010; Avugos et al., 2013; Bar-Eli et al., 2006) and with special focuses on the impact of high-pressure game situations (Madden et al., 1990, 1995; Goldman and Rao, 2012; Zuccolotto et al., 2018),

5. monitoring playing patterns with reference to roles (Sampaio et al., 2006), also with the aim of defining new playing positions (Alagappan, 2012; Bianchi et al., 2017),

6. designing the kinetics of players' body movements with respect to shooting efficiency, timing and visual control on the field (Miller

and Bartlett, 1996; Okubo and Hubbard, 2006; de Oliveira et al., 2006; Aglioti et al., 2008),

7. depicting the players' movements, pathways, trajectories from the in-bounds pass to the basket and the network of passing actions (Fujimura and Sugihara, 2005; Perše et al., 2009; Skinner, 2010; Therón and Casares, 2010; Bourbousson et al., 2010b,c; Passos et al., 2011; Lamas et al., 2011; Piette et al., 2011; Fewell et al., 2012; Travassos et al., 2012; Shortridge et al., 2014; Ante et al., 2014; Clemente et al., 2015; Gudmundsson and Horton, 2016; Metulini et al., 2017a,b; Wu and Bornn, 2018; Bornn et al., 2017; Miller and Bornn, 2017; Metulini et al., 2018), the flow of events and the connected functional decisions which have to be taken by players during the game (Araújo et al., 2006, 2009; Araújo and Esteves, 2010; Perica et al., 2011; Araújo and Davids, 2016),

8. studying teams' tactics and identifying optimal game strategies (Annis, 2006; Zhang et al., 2013; Skinner and Goldman, 2017),

9. investigating the existence of possible referee biases (Noecker and Roback, 2012),

10. measuring psychological latent variables and their association to performance (Meyers and Schleser, 1980; Weiss and Friedrichs, 1986; Seifriz et al., 1992; Taylor, 1987; Maddi and Hess, 1992; Dirks, 2000; Heuzé et al., 2006; Bourbousson et al., 2010a).

Although composed of just short of one hundred references, this literature review is far from being complete, as the number of papers published on these topics is really huge.

In addition, we did not cite whole research areas that are not fully relevant to the aims of this book. For example, a wide literature covers medical issues related to the epidemiology of basketball injuries, physical, anthropometric and physiological attributes of players, association between performance and hematological parameters or other vitals (heart rate, blood pressure, aerobic capacity, etc.), special training techniques to stimulate muscle strength, jumping ability and physical fitness in general. Lastly, it is worth mentioning all the literature dealing with scheduling problems, proposing approaches to solve the myriad of conflicting requirements and preferences implied by the creation of a suitable schedule.

A comprehensive literature review can be found in Passos et al. (2016), where a whole chapter is devoted to previous research on basketball data science (other chapters are concerned with soccer, other football codes, ice hockey, handball, volleyball). Other reviews have been drawn on specific topics, *e.g.*, on the home advantage (Courneya and Carron, 1992), on knee injury patterns (Arendt and Dick, 1995) or other medical matters, but only sometimes focused just on basketball and more often addressing sports in general or, at most, team sports.

Despite this already wide landscape, the range of possible questions that may be answered by Data Science is going to grow, thanks to the availability of large sets of data (even from the lowest-level tournaments) and to the increasing computational power. As a matter of fact, from a methodological point of view, the huge amount of raw data available on the actions and strategies, combined with the absence of a sound theory explaining the relationships between the many involved variables, make these questions a challenging heat for Data Scientists.

1.3 HOW THE BOOK IS STRUCTURED

This book is structured in three parts: the first one introduces the fundamentals of Basketball Analytics (literature review, data description and basic methods), while the second part is devoted to advanced methods of Basketball Data Science. All the analyses presented throughout these two parts are carried out by means of `BasketballAnalyzeR`, an R package purposely designed with this book. Relevant commands and codes are made available to readers, in order to allow everybody to reproduce the presented analyses. The third part is a technical appendix focused on `BasketballAnalyzeR`.

In Chapter 2, after a brief description of the main kinds of basketball data and the datasets that are used for the case studies presented throughout the book, some descriptive statistical tools are examined and discussed. In detail, we touch on pace, Offensive and Defensive Ratings, Dean Olivers' Four Factors, several kinds of plots such as bar-line, radial, bubble and scatterplots, variability and inequality analysis, shots charts with scoring percentages from different areas of the court.

The second part of the book, devoted to advanced methods, follows the categorization into the three major issues outlined in Section 1.1.1: discovering patterns in data, finding groups in data, and modeling relationships in data.

More specifically, Chapter 3 deals with discovering patterns in data and, after addressing some basic concepts about associations between variables and linear correlation, draws attention to displaying maps of individual cases according to their similarity, visualizing network relationships among players, estimating the shots' density with respect to some exogenous variables such as, for example, time played, play length and shot distance.

Chapter 4 is devoted to finding groups in data, thus, technically speaking, to cluster analysis. Two main methods of cluster analysis are examined, namely the k-means and the hierarchical algorithm, and some case studies are discussed. Cluster analysis can be used in basketball studies with several aims; for example, for identifying groups of players sharing similar performance indicators (also despite different roles) or groups of matches, according to game final score differences, in order to find regularities in those indicators that most distinguish between winning and losing performances within matches, or groups of teams with similar achievements during a championship, or groups of shots with similar characteristics, or groups of time instants during a match, according to the players' behavior and strategies, etc.

Finally, Chapter 5 is concerned with statistical modeling. Linear models and nonparametric regression are specifically addressed, with some case studies with a bivariate setting. Some specific issues are also considered, namely the problem of estimating the expected points and the scoring probability of a shot as a function of some specific variables such as for example, time played, play length and shot distance.

In each chapter of the second part, a FOCUS Section is devoted to put under the spotlight more complex case studies taken from the scientific literature and summarize their main findings pruning down from the most technical details. In Chapter 3, the FOCUS Section deals with using machine learning algorithms to assess the effects of high-pressure game situations on the scoring probability; in Chapter 4, the problem of classifying new playing styles into new roles and positions is handled with advanced cluster analysis methods based on unsupervised neural networks; finally, the FOCUS Section of Chapter 5 is concerned with the use of sensor data recording players' positions on the court at high frequency in order to determine the surface area dynamics and their effects on the team performance.

Whatever the modeling technique or application, Data Science needs programming tools for data preparation, analysis, visualization, and modeling. A lot of options are available to this aim, and Data Scientists' favorite alternatives are open-source tools (Chen et al., 2007; Miller, 2015). In this book, all the presented applications will be developed using

R, a language and environment for statistical computing and graphics (Ihaka and Gentleman, 1996; R Development Core Team, 2008), consisting of more than 13,000 packages.

The third part of this book consists of a technical appendix written by Marco Sandri about `BasketballAnalyzeR`. Chapter 6 describes its main features and data preparation. Finally, some advanced issues on customizing plots and building interactive graphics are discussed. Further explanations can be found on the functions' help pages of the package, while codes, interactive graphs, news about the package, possible updates, and discussions about data preparation are made available in a webpage specifically devoted to it and addressed to the readers of this book, at

`https://bdsports.unibs.it/basketballanalyzer/`

This webpage is constantly updated, so it has to be considered a reference point by all the users of `BasketballAnalizeR`, who can also find there contact information for any question they may want to ask the developers. In addition, the readers of this book will find there the supplementary material of this book, where the R code is downloadable allowing to reproduce all the case studies presented throughout the book.

GLOSSARY

Data Science: Interdisciplinary science dealing with methods, tools, processes, algorithms and systems aimed to extract knowledge or insights from data in various forms, either structured or unstructured, in small or big amounts.

Open Source: The open-source venture in software development is a response of the scientific community to the growing costs and limitations of proprietary codes. The open-source philosophy promotes a decentralized model for software development, encouraging a collaborative effort of cooperation among peers, who have free universal access to source codes and documentation and share changes/improvements within the community.

CHAPTER 2

Data and Basic Statistical Analyses

B ASIC METHODS for the analysis of sports data typically consist of indexes and charts. In the case of basketball, as already pointed out in Section 1.2, a wide set of such analytics is available, starting from the pioneering work of Dean Oliver. The main part of these analytics can be easily computed and interpreted by practitioners and, apart from being slightly off topic for a book on Data Science, this subject is too broad to be discussed in a few pages. Among the countless examples of such analytics, we will limit ourselves to cite some hugely common analytics based on the concepts of possession and pace (Oliver, 2004; Kubatko et al., 2007) in Section 2.2.1. The other basic Statistics this chapter will be dealing with are tools of descriptive Statistics applied to basketball data: bar plots (Section 2.2.2) and radial plots (Section 2.2.3) built using game variables, percentages or other standardized statistics, scatter plots of two selected variables (Section 2.2.4), bubble plots able to represent several features of teams or players in a unique graph (Section 2.2.5), variability (Section 2.2.6) and inequality (Section 2.2.7) indexes and graphs, shot charts with the court split into sectors colored according to a selected game variable and annotated with scoring percentages (Section 2.2.8). All the analyses are carried out using the language R (R Development Core Team, 2008), and the package BasketballAnalyzeR (see Chapter 6). The R code provided throughout the book has been checked under R-3.5.3 for Windows (64 bit) and for Mac OS X 10.11 and can be fully downloaded in the webpage mentioned few rows below. General references for learning R include Chambers (2008), Matloff (2011) and Wickham (2014). The package must be preliminarily installed following the instructions at

https://bdsports.unibs.it/basketballanalyzer/

and then loaded by typing

```
> library(BasketballAnalyzeR)
```

The introduction on basic statistical analyses shall be preceded by a brief description of the most common types of basketball data.

2.1 BASKETBALL DATA

Data are the life blood of Data Science; therefore, the procedures for obtaining and organizing datasets must be structured and validated in order to guarantee quality, understood as exhaustiveness (presence of the variables of interest to carry out the required analyses), accuracy (absence or minimization of errors), completeness (absence or possible treatment of missing data), consistency (presence of a large number of observations, necessary for the robustness of statistical analyses), accessibility (possibility to find them easily and quickly, in searchable form), and timeliness. A complete discussion on the different approaches to assessing data quality can be found in Kenett and Shmueli (2016). Another important issue about data is context (Cobb and Moore, 1997), that is all the additional information (definitions, methods, technologies, conditions, environment, etc.) surrounding data and whichever way affecting how we might interpret them. Data without context are just numbers.

Data are not just numbers, they are numbers with a context.

George W. Cobb and David S. Moore
Distinguished professors of Mathematics and Statistics

Data can be obtained through multiple channels and sources, such as National and International Federations, sporting organizations, professional societies, sport-related associations and other special interest sources (see Kubatko et al., 2007 and Schumaker et al., 2010 for a brief review). The web is a massive store of data. Data may be on payment or freely available on specific websites; therefore, the collecting procedures are various. Access to open source channels sometimes requires high-level computer skills, such as knowledge of web scraping procedures. In addition, datasets can have different characteristics from the point of view of size, variety and speed of update: we move from small datasets to big data, so that flexible, scalable and distributed systems are necessary. Variety, on its part, requires to manage relational databases and data warehousing tools (Golfarelli and Rizzi, 2009) in order to accommodate and properly handle data with a broad spectrum of different features: traditional datasets arranged on two-dimensional grids with cases on the rows and variables on the columns, multidimensional data cubes, unstructured text data, pixels from sensors and cameras, data from wearables, mobile phones, tablets, field-of-play coordinates, geocodes with latitude and longitude, timestamps showing time to the nearest millisecond, etc.

To attempt a rough classification of data in macro-categories, we can distinguish:

- **Data recorded manually**, with or without technological tools for annotation. This category includes the basic statistics from box scores, notational analysis data, play-by-play (event-log) data, reports filled by technical experts and coaches during training sessions, opinions and experts' evaluations that can be combined with measurement data.

- **Data detected by technological devices**. Increasingly, technology enters both the training and the games, making large amounts of data available. Examples are the data recorded by GPS sensors or other player tracking systems, which detect the positions of the players on the court at very short time intervals (milliseconds), the video data coming from cameras, the platforms and all the wearable technologies that detect postures, body movements, vitals such as heartbeat and blood pressure.

- **Data from psychometric questionnaires** administered to athletes, aimed at the measurement of attitudes and personality traits (group dynamics, interpersonal relations, social-cognitive processes, leadership, mental toughness, personality, coping strategies, etc.).

- **Other data**. In this residual category, converge all the different and heterogeneous data classes that can integrate the analysis from different points of view, such as - without pretension of exhaustive-ness - the market analysis data, the textual data obtained by query-ing the Social Networks (which can serve for example to measure the sentiment of fans), data from Google Trends and other tools able to monitor online searches and popularity of hashtags.

Once gathered, raw data are often unstructured, messy, and some-times missing. It is the Data Scientist's duty to organize, clean, and complete them, pursuing the purpose of data quality. This preliminary work needs proper skills and tools, which will not be discussed in this book. Relevant readings in this context are Hernández and Stolfo (1998), Rahm and Do (2000), Allison (2001), Kim et al. (2003) and Little and Rubin (2014).

In this book, examples and case studies will be developed using dif-ferent datasets, all available in the R package `BasketballAnalyzeR`, as can be seen from

```
> data(package="BasketballAnalyzeR")
```

In detail, we will consider three datasets of box scores, one play-by-play dataset and a supplementary data frame containing additional qualita-tive information about the teams:

1. **Teams' box scores.** In this data frame, called `Tbox`, the cases (rows) are the analyzed teams and the variables (columns) are referred to the team achievements in the considered games. Vari-ables are listed in Table 2.1.

2. **Opponents' box scores.** In this data frame, called `Obox`, the cases (rows) are the analyzed teams and the variables (columns) are referred to the achievements of the opponents of each team in the considered games. Variables are listed in Table 2.1.

3. **Players' box scores.** In this data frame, called `Pbox`, the cases (rows) are the analyzed players and the variables (columns) are referred to the individual achievements in the considered games. Variables are listed in Table 2.1.

4. **Play-by-play data.** In this data frame, called `PbP.BDB`, the cases (rows) are the events occurred during the analyzed games and the variables (columns) are descriptions of the events in terms of type, time, players involved, score, area of the court. Variables are listed in Table 2.2.

TABLE 2.1   Variables of data frames **Tbox**, **Obox**, **Pbox** and **Tadd**.

Variable	Description	Tbox	Obox	Pbox	Tadd
Team	Analyzed team (long name)	×	×	×	×
team	Analyzed team (short name)				×
Conference	Conference				×
Division	Division				×
Rank	Rank (end season)				×
Playoff	Playoff qualification (Yes or No)				×
Player	Analyzed player			×	
GP	Games Played	×	×	×	
MIN	Minutes Played	×	×	×	
PTS	Points Made	×	×	×	
W	Games won	×	×		
L	Games lost	×	×		
P2M	2-Point Field Goals (Made)	×	×	×	
P2A	2-Point Field Goals (Attempted)	×	×	×	
P2p	2-Point Field Goals (Percentage)	×	×	×	
P3M	3-Point Field Goals (Made)	×	×	×	
P3A	3-Point Field Goals (Attempted)	×	×	×	
P3p	3-Point Field Goals (Percentage)	×	×	×	
FTM	Free Throws (Made)	×	×	×	
FTA	Free Throws (Attempted)	×	×	×	
FTp	Free Throws (Percentage)	×	×	×	
OREB	Offensive Rebounds	×	×	×	
DREB	Defensive Rebounds	×	×	×	
AST	Assists	×	×	×	
TOV	Turnovers	×	×	×	
STL	Steals	×	×	×	
BLK	Blocks	×	×	×	
PF	Personal Fouls	×	×	×	
PM	Plus/Minus	×	×	×	

TABLE 2.2   Variables of data frame PbP.BDB.

Variable	Description
game_id	Identification code for the game
data_set	Season: years and type (Regular or Playoffs)
date	Date of the game
a1 …a5; h1 …h5	Five players on the court (away team; home team)
period	Quarter (≥ 5: over-time)
away_score; home_score	Score of the away/home team
remaining_time	Time left in the quarter (h:mm:ss)
elapsed	Time played in the quarter (h:mm:ss)
play_length	Time since the immediately preceding event (h:mm:ss)
play_id	Identification code for the play
team	Team responsible for the event
event_type	Type of event
assist	Player who made the assist
away; home	Players for the jump ball
block	Player who blocked the shot
entered; left	Player who entered/left the court
num	Sequence number of the free throw
opponent	Player who made the foul
outof	Number of free throws accorded
player	Player responsible for the event
points	Scored points
possession	Player whom the jump ball is tipped to
reason	Reason of the turnover
result	Result of the shot (made or missed)
steal	Player who stole the ball
type	Type of play
shot_distance	Field shots: distance from the basket
original_x ; original_y; converted_x ; converted_y	coordinates of the shooting player original: tracking coordinate system half court, (0,0) center of the basket converted: coordinates in feet full court, (0,0) bottom-left corner
description	Textual description of the event

5. **Additional information**. In this data frame, called `Tadd`, the cases (rows) are the analyzed teams and the variables (columns) are qualitative information such as Conference, Division, final rank, qualification for Playoffs. Variables are listed in Table 2.1.

All the data frames contained in the package relate to the whole regular season of the NBA championship 2017/2018 (82 games); box scores and additional information are about all the teams and players, play-by-play data are relative to the 82 games played by the Champions, Golden State Warriors, during the regular season. Play-by-play data have been kindly made available by BigDataBall (www.bigdataball.com), a data provider which leverages computer-vision technologies to enrich and extend sports datasets with lots of unique metrics. Since its establishment, BigDataBall has also supported many academic studies and is referred to as a reliable source of validated and verified stats for NBA, MLB, NFL and WNBA. The R functions of `BasketballAnalyzeR` requiring play-by-play data as input need a data frame with some additional variables with respect to `PbP.BDB`. It can be obtained by typing

```
> PbP <- PbPmanipulation(PbP.BDB)
```

where `PbPmanipulation` is a function that adapts the standard file supplied by BigDataBall to the format required by `BasketballAnalyzeR`. The resulting data frame `PbP` is composed of the same variables of `PbP.BDB` (when necessary, coerced from one data type to another, *e.g.*, from factor to numeric) plus five additional variables listed in Table 2.3. In the following, we will use the data frame `PbP` in all the analyses involving play-by-play data.

Throughout the book, we will show how the functions of `BasketballAnalyzeR` work with the above-mentioned datasets. Researchers can replicate the described analyses (or perform new ones)

TABLE 2.3   Additional variables of data frame `PbP`.

Variable	Description
`periodTime`	Time played in the quarter (in seconds)
`totalTime`	Time played in the match (in seconds)
`playlength`	Time since the immediately preceding event (in seconds)
`ShotType`	Type of shot (FT, 2P, 3P)
`oppTeam`	Name of the opponent team

on their own data, provided some rules about data preparation are followed. The easiest and safest way for a correct replication on an arbitrary data set of the analyses considered in this book is to build data frames with the same structure of the Tbox, Obox, Pbox, Tadd and PbP datasets. Further details on this issue and examples will be discussed in Chapter 6, Section 6.2.

2.2 BASIC STATISTICAL ANALYSES

In this section some basic tools will be discussed with examples. Virtually, the analytics that will be described can be computed on any team or player for which box scores are available and on any play-by-play dataset, considering a single match or more games grouped together, in any moment of a championship or a tournament.

2.2.1 Pace, Ratings, Four Factors

This section is based on Kubatko et al. (2007), the first paper that brought the generally accepted mainstream of basketball analytics to a scientific journal, thereby establishing a common starting point for data science research in this field. In detail, we address the concepts of possession and pace, Offensive/Defensive Ratings and the notorious Four Factors. We denote variables with the same names of the R data frame columns in Table 2.1.

Following Kubatko et al. (2007), possessions are computed as

$$POSS = (P2A + P3A) + 0.44 \times FTA - OREB + TOV \qquad (2.1)$$

and the strictly related statistic accounting for the pace of the game is given by

$$PACE = 5 \times POSS/MIN \qquad (2.2)$$

where MIN are the total minutes played by all the players. Per-possession efficiency is measured as the points scored or allowed per 100 possessions, called Offensive ($ORtg$) and Defensive ($DRtg$) Rating, respectively:

$$ORtg = PTS_T/POSS_T \qquad (2.3)$$

$$DRtg = PTS_O/POSS_O \qquad (2.4)$$

where the subscripts T and O refer to whether the statistic is computed on the analyzed team or the opponent(s). Finally, the Four Factors -

TABLE 2.4   Four-Factor formulas (for offense and defense).

Factor	Offense	Defense
$eFG\%$	$\dfrac{P2M_T + 1.5 \times P3M_T}{P2A_T + P3A_T}$	$\dfrac{P2M_O + 1.5 \times P3M_O}{P2A_O + P3A_O}$
TO Ratio	$\dfrac{TOV_T}{POSS_T}$	$\dfrac{TOV_O}{POSS_O}$
$REB\%$	$\dfrac{OREB_T}{OREB_T + DREB_O}$	$\dfrac{DREB_T}{OREB_O + DREB_T}$
FT Rate	$\dfrac{FTM_T}{P2A_T + P3A_T}$	$\dfrac{FTM_O}{P2A_O + P3A_O}$

Effective Field Goal Percentage ($eFG\%$), Turnovers per possession (TO Ratio), Rebounding percentages ($REB\%$), Free Throw Rate (FT Rate) - for each the offense and the defense are computed as in Table 2.4.

The function `fourfactors` of the package `BasketballAnalyzeR`, easily computes and graphically represents all these indexes for one or more selected teams. For example, if we wish to limit attention to the Conference finalists (Boston Celtics, Cleveland Cavaliers, Golden State Warriors, Houston Rockets), we use the following code

```
> tm <- c("BOS","CLE","GSW","HOU")
> selTeams <- which(Tadd$team %in% tm)
> FF.sel <- fourfactors(Tbox[selTeams,], Obox[selTeams,])
```

where BOS, CLE, GSW, and HOU are the short names of the four selected teams. The R object FF (of class `fourfactors`) contains a data frame with Possessions, Pace, Offensive/Defensive Ratings and Four Factors as columns. A `plot` method is available for this class, so that the command

```
> plot(FF.sel)
```

returns the graphs in Figure 2.1, which allows some important remarks about the analyzed teams:

- **Pace**: the pace of the games increases moving from Boston Celtics to Houston Rockets, Cleveland Cavaliers and Golden State Warriors.

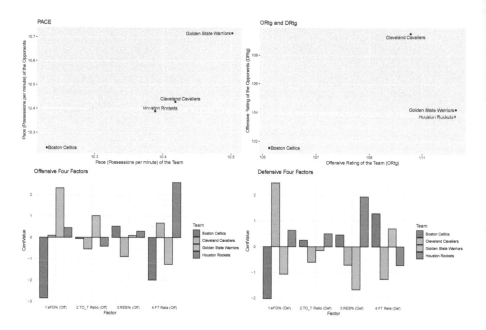

Figure 2.1 Pace, Offensive/Defensive Ratings and Four Factors (differences between the team and the average of the considered teams) - Conference finalists 2017/2018.

- **Offensive/Defensive Ratings**: The Boston Celtics and Cleveland Cavaliers have respectively low and high values for both $ORtg$ and $DRtg$; the Golden State Warriors and Houston Rockets have the best performance from this perspective, as they have high $ORtg$ and a relatively low $DRtg$.

- **Offensive/Defensive Four Factors**: The bars represent, for each team and each Factor, the difference between the team value and the average of the four analyzed teams. The positive or negative value of this difference and the height of the corresponding bar give a clear idea of strengths and weaknesses of the teams with respect to each other.

An R object of class `list` containing the four graphs (generated by the `ggplot2` library Wickham, 2016) of Figure 2.1 can be created (and customized). Suppose this time we want to analyze all the teams: we first generate the object `FF` of class `fourfactors`, then the object `listPlots` of class `list`

```
> FF <- fourfactors(Tbox,Obox)
> listPlots <- plot(FF)
```

Now the four graphs can be plotted separately or can be arranged in a single plot, using the R package **gridExtra** (Auguie, 2017). For example, the graphs of Pace and Offensive/Defensive Ratings of all the teams can be displayed together as follows (Figure 2.2):

```
> library(gridExtra)
> grid.arrange(grobs=listPlots[1:2], ncol=1)
```

2.2.2 Bar-line plots

A useful graphical tool, often used to visualize comparisons among teams or players according to some selected statistics, is the bar-line plot. In this graph, for each analyzed team/player, a bar is drawn, whose height is determined by one ore more (in the case of stacked bars) variables. The bars can be ordered according to a selected variable. Furthermore, relevant information can be added to the chart by means of a line, whose scale can be read on another axis, conventionally put on the right. There are no crucial statistical issues in composing this graph, what only matters is the right choice of the variables, in order to obtain an insightful representation of reality. Some examples follow on how the function **barline** can be used to this extent in a very flexible way.

The graph in the top panel of Figure 2.3 is obtained from the following code lines

```
> X <- data.frame(Tbox, PTS.O=Obox$PTS, TOV.O=Obox$TOV,
                  CONF=Tadd$Conference)
> XW <- subset(X, CONF=="W")
> labs <- c("Steals","Blocks","Defensive Rebounds")
> barline(data=XW, id="Team", bars=c("STL","BLK","DREB"),
         line="TOV.O", order.by="PTS.O", labels.bars=labs)
```

and represents the main defensive statistics of the Western Conference NBA teams: steals, blocks and defensive rebounds for the bars, ordered (in decreasing order) according to the points scored by the opponents (variable $PTS.Opp$); the gray line represents turnovers of the opponents (whose scale is on the right vertical axis). The graph does not highlight any evident relationship between the defensive statistics and the points scored by the team's opponents or their turnovers. In fact, the teams

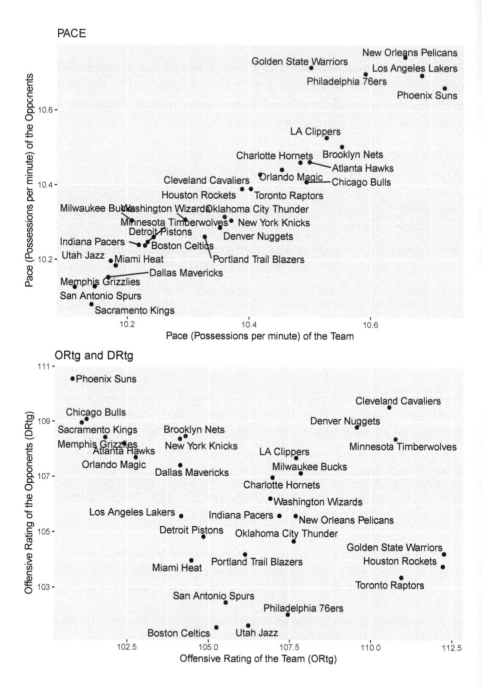

Figure 2.2 Pace and Offensive/Defensive Ratings - NBA teams 2017/2018.

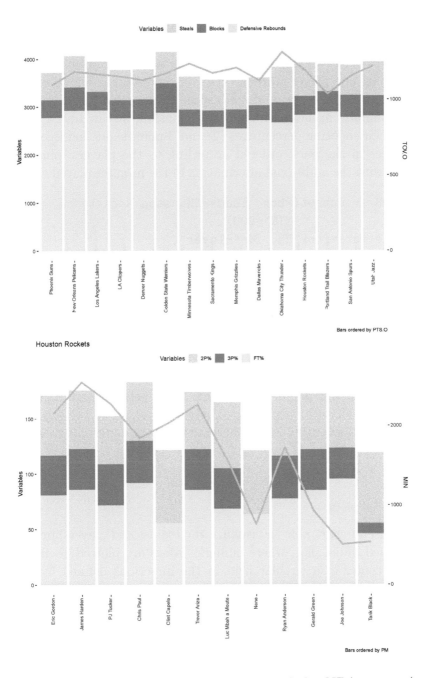

Figure 2.3 Bar-line plots: defensive statistics of the NBA teams (top: TOV.O = turnovers of the opponents, PTS.O = points scored by the opponents) - shooting percentages and minutes played by Houston Rockets players (bottom).

on the left bars (those with the highest values of points scored by the opponents, *e.g.*, Phoenix Suns $PTS.Opp = 9290$, New Orleans Pelicans $PTS.Opp = 9054$) are not necessarily those with the lowest bars.

Alternatively, player statistics can be represented. Suppose we are interested in the players of the Houston Rockets who have played at least 500 minutes in the considered championship, we can draw a bar-line plot for their shooting percentages (2- and 3-point shots and free throws percentages in the bars, ordered according to the players' plus-minus statistic) with the additional information of the minutes played (line):

```
> Pbox.HR <- subset(Pbox, Team=="Houston Rockets" &
              MIN>=500)
> barline(data=Pbox.HR, id="Player",
          bars=c("P2p","P3p","FTp"), line="MIN",
          order.by="PM", labels.bars=c("2P%","3P%","FT%"),
          title="Houston Rockets")
```

as displayed in the bottom panel of Figure 2.3. This graph shows that the players on the left bars (those with the highest values of plus-minus statistic, *e.g.*, Eric Gordon $PM = 589$, James Harden $PM = 525$) tend to also have the highest minutes played (respectively, $MIN = 2154$ and $MIN = 2551$), but not necessarily the best shooting performance, as there are players with shooting percentages as good as the top ranked players, but considerably lower plus-minus statistic and minutes played (*e.g.*, Gerald Green $PM = 93$ and $MIN = 929$, Joe Johnson $PM = 18$ and $MIN = 505$).

2.2.3   Radial plots

Another useful chart able to visualize the teams' and players' profiles is the so-called radial plot, where numeric values are plotted as distances from the center of a circular field in the directions defined by angles obtained dividing the 360-degree angle as many times as the number of considered variables. Joining the points representing the numerical values of the variables, a polygon is obtained, visually describing the profile of the considered subject (team or player). We must mention here that some researchers raise criticism about this kind of graph. According to them, it makes comparisons about data points difficult because it draws the eye to the area, rather than the distance of each point from the center. In addition, the area is determined by the arbitrary choice of the sequence of variables and there is a graphical over-emphasis on high numbers. However, it can be very useful when an immediate comparison of different profiles is necessary, as, for example, in Cluster Analysis (see Chapter 4). We recommend to be wary in interpreting it, as shown in the following examples.

The function `radialprofile` displays radial plots of teams or players, with a flexible choice of variables and the possibility to standardize them in order to enhance differences among the analyzed cases. It is built starting from the R function `CreateRadialPlot` by Vickers (2006), freely downloadable from the web.

For example, we may be interested in comparing profiles of nine point guards, namely Russell Westbrook, Stephen Curry, Chris Paul, Kyrie Irving, Damian Lillard, Kyle Lowry, John Wall, Rajon Rondo and Kemba Walker according to 2- and 3-point shots made, free throws made, total rebounds (offensive and defensive), assists, steals and blocks (per minute played). The code lines are

```
> Pbox.PG <- subset(Pbox, Player=="Russell Westbrook" |
                    Player=="Stephen Curry" |
                    Player=="Chris Paul" |
                    Player=="Kyrie Irving" |
                    Player=="Damian Lillard" |
                    Player=="Kyle Lowry" |
                    Player=="John Wall" |
                    Player=="Rajon Rondo" |
                    Player=="Kemba Walker")
> attach(Pbox.PG)
> X <- data.frame(P2M, P3M, FTM, REB=OREB+DREB, AST,
                    STL, BLK)/MIN
> detach(Pbox.PG)
> radialprofile(data=X, title=Pbox.PG$Player, std=FALSE)
```

We obtain the graph of Figure 2.4, where we note that, although each player has his own features, there are some similar profiles (Irving, Walker and Lillard, Rondo and Wall, Curry and Paul). In spite of its apparent straightforwardness, we have to pay a great deal of attention while setting up this graph. In fact, the axes have all the same scale, ranging from the absolute minimum to the absolute maximum (with the dashed blue line in the midpoint between them) of all the analyzed variables. For this reason, we should select game features having about the same scale; otherwise, some variables would be shrinked on axes set with a scale too wide for them. This would make us unable to assess

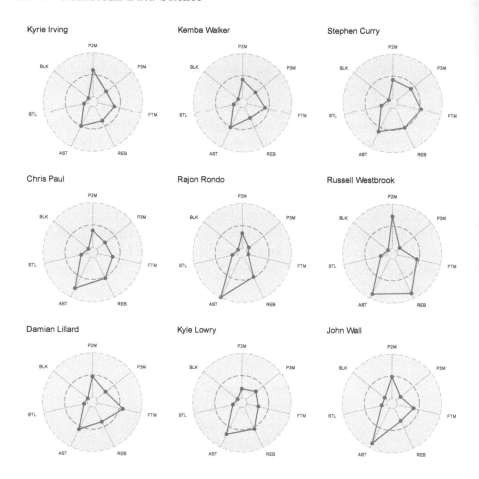

Figure 2.4 Radial plots of nine selected point guards, non-standardized variables. Dashed blue line: midpoint between minimum and maximum.

differences among players with respect to those variables. In some small way, this happens in the graph of Figure 2.4, where the axes range from min(X) ($\simeq 0.005$) to max(X) ($\simeq 0.313$). For example, variable STL (steals per minute played) ranges from about 0.029 to 0.052 and differences among players on this aspect - although considerable - are not visually detectable in the graph.

In this situation, it is recommendable to complement the analysis with the graph obtained by setting the option std=TRUE, thanks to which the analyzed variables are preliminarily standardized. We recall that the standardization of a variable X with mean μ_X and standard deviation

σ_X is a linear transformation generating a new variable Z given by

$$Z = \frac{X - \mu_X}{\sigma_X}$$

with mean $\mu_Z = 0$ and standard deviation $\sigma_Z = 1$. In other words, in the graph obtained with std=TRUE, all the variables are transformed into the same scale, and the dashed blue line is drawn at the zero level, corresponding, for each variable, to its average. The points in the profiles can then be interpreted as being, for each variable, above or below the average of the analyzed players. In Figure 2.5 we see how the profiles

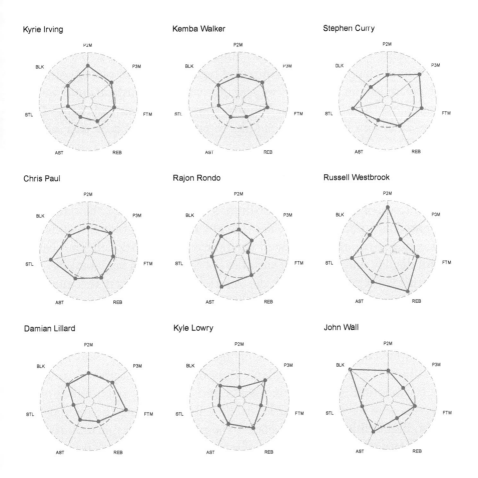

Figure 2.5 Radial plots of nine selected point guards, standardized variables. Dashed blue line: zero (average of each variable).

change as a result of the standardization. Several remarks are possible: there are point guards who distinguish themselves for a particularly high (Curry and Lowry) or low (Rondo, Westbrook and Wall) number of 3-point shots made. The same can be said for 2-point shots (high: Irving and Westbrook; low: Rondo and Lowry), steals (high: Curry, Paul and Westbrook; low: Irving, Walker, Lillard and Lowry), and all the other analyzed variables. Jointly considering all the variables, the most outstanding profile is that of Westbrook, having all the variables above average, except 3-point shots made and blocks (but note that the blocks' average is boosted by the impressive performance of Wall, in the top 20% of the entire league that year for this variable - even more remarkable considering his size and that he plays the point guard position). Two last warnings have to be given about the standardized version of this graph: firstly, when we interpret a value as being above or below average, of course that average is computed on the analyzed players and not on the entire league; secondly, we should avoid standardization if we are analyzing less than 5-6 players. Finally, consider that in `radialprofile` the axes scale can be adjusted thanks to the argument `min.mid.max`.

2.2.4 Scatter plots

A scatter plot is a graph displaying values for two variables observed on a set of data using Cartesian coordinates. Each subject of the dataset is represented as a point having the value of the two variables as x-axis and y-axis coordinates, respectively. If the points are color-coded, the scatter plot can display one additional variable (as the coloring variable). Scatter plots can give a rough idea of the relationships between the analyzed variables, suggest various kinds of associations and highlight anomalous cases.

The function `scatterplot` allows to display scatter plots with a set of options for color-coding, highlighting selected subjects, and zooming into a subset of the Cartesian plane. Suppose we are interested to investigate the relationship between assists and turnovers per minute of all the players who have played at least 500 minutes during the regular season, also highlighting, by color codes, the points scored per minute. The code lines producing the graph in the top panel of Figure 2.6 are

```
> Pbox.sel <- subset(Pbox, MIN>= 500)
> attach(Pbox.sel)
> X <- data.frame(AST, TOV, PTS)/MIN
> detach(Pbox.sel)
```

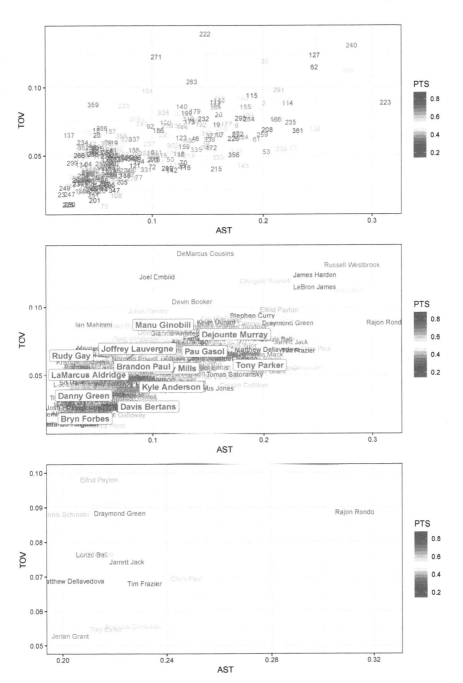

Figure 2.6 Scatter plots of (per minute) assists versus turnovers with points colored according to the points scored per minute. Points are plotted by case numbers (top); by players' names, with selected players in evidence (middle); in a selected portion of the plane (bottom).

```
> mypal <- colorRampPalette(c("blue","yellow","red"))
> scatterplot(X, data.var=c("AST","TOV"), z.var="PTS",
              labels=1:nrow(X), palette=mypal)
```

Note that for the option `palette` the common default setting (*e.g.*, `rainbow`, `terrain.colors`, ...) can be used instead of specifying the customized palette `mypal`.

Looking at the graph, we note that the number of turnovers tends to increase with increasing assist values. This reflects a clear insight: the more a player moves the ball, the more likely he turns it over. It is worth noting some cases of players with high turnovers with respect to assists (case numbers 271, 283, 222), with high assists with respect to turnovers (case numbers 132, 82, 223), with both high assists and turnovers (case numbers 62, 127, 240, 357). All these outstanding cases, except 223, exhibit high values of scored points.

In order to plot players' names instead of case numbers, we add the option `labels=Pbox.sel$Player` in the function `scatterplot`. In addition, the positioning in the scatter of a subset of players can be highlighted thanks to the option `subset`. Suppose we are interested in the players of the San Antonio Spurs, the following code lines produce the graph in the middle panel of Figure 2.6.

```
> SAS <- which(Pbox.sel$Team=="San Antonio Spurs")
> scatterplot(X, data.var=c("AST","TOV"), z.var="PTS",
              labels=Pbox.sel$Player, palette=mypal,
              subset=SAS)
```

Finally, to zoom into a subset of the Cartesian plane, say $[0.20, 0.325] \times [0.05, 0.10]$ (corresponding to players who exhibit high assist values associated to moderately low turnovers), we add the option[1] `zoom=c(0.20,0.325,0.05,0.10)` and obtain the graph in the bottom panel of Figure 2.6.

When three or more variables are considered, the function `scatterplot` allows to plot scatter plot matrices, as will be described later, in Chapter 3, Section 3.2. Several additional options are also available, that will be considered when appropriate (see Chapter 3, Section 3.5.3 and Chapter 5, Section 5.2).

---

[1]Note that when the argument `zoom` is used, the function generates a warning message for the impossibility to plot the subjects outside the selected area, which of course can be ignored when the aim is to zoom.

2.2.5   Bubble plots

A bubble plot is a scatter plot where individual cases (teams or players) are plotted in the plane by means of bubbles instead of points. The size and color of the bubble are defined according to two additional variables, so that the bubbles' scatter includes information on four features. The function **bubbleplot** easily allows to build such a graph, with variables selected by the researcher. A bubble plot of the teams with variables given by the shooting percentages (2- and 3-point shots on the x- and y-axes, respectively, free throws on a red-blue color scale) and the total number of attempted shots (the size of the bubbles) can be obtained with the following code lines (Figure 2.7):

```
> attach(Tbox)
> X <- data.frame(T=Team, P2p, P3p, FTp, AS=P2A+P3A+FTA)
```

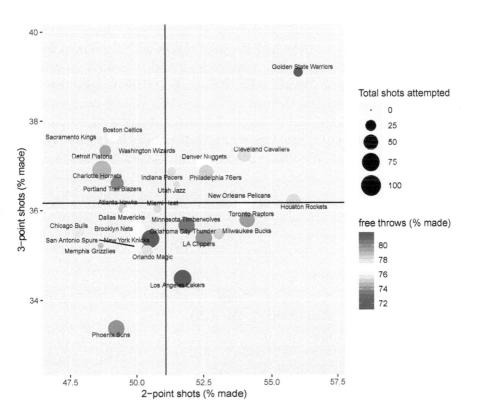

Figure 2.7   Bubble plot of the teams according to shooting percentages and total shots attempted.

```
> detach(Tbox)
> labs <- c("2-point shots (% made)",
            "3-point shots (% made)",
            "free throws (% made)",
            "Total shots attempted")
> bubbleplot(X, id="T", x="P2p", y="P3p", col="FTp",
            size="AS", labels=labs)
```

Note that, in order to improve readability, the bubble size is rescaled between 0 and 100 (corresponding to the minimum and the maximum number of total shots attempted, respectively). This rescaling can be disabled setting the argument `scale.size` to `FALSE` (it is `TRUE` by default). Looking at the graph, we immediately note the outstanding location of Golden State Warriors, who exhibit the highest shooting percentages but, surprisingly, a low number of attempted shots.

Another example of bubble plot can be obtained representing the players of two teams (say, Golden State Warriors and Cleveland Cavaliers), focusing just on those who have played at least 500 minutes in the championship, according to their statistics related to defense: defensive rebounds, steals and blocks per minute played. Let the bubble size represent the total minutes on the court (Figure 2.8). We set `scale=FALSE` in order to represent the real number of minutes played and we also use the argument `text.col`, thanks to which the bubbles' labels are colored according to the team. With `text.size` we are allowed to customize the labels' size.

```
> Pbox.GSW.CC <- subset(Pbox,
                        (Team=="Golden State Warriors" |
                        Team =="Cleveland Cavaliers") &
                        MIN>=500)
> attach(Pbox.GSW.CC)
> X <- data.frame(ID=Player, Team, V1=DREB/MIN, V2=STL/MIN,
                V3=BLK/MIN, V4=MIN)
> detach(Pbox.GSW.CC)
> labs <- c("Defensive Rebounds","Steals","Blocks",
            "Total minutes played")
> bubbleplot(X, id="ID", x="V1", y="V2", col="V3",
            size="V4", text.col="Team", labels=labs,
            title="GSW and CC during the regular season",
            text.legend=TRUE, text.size=3.5, scale=FALSE)
```

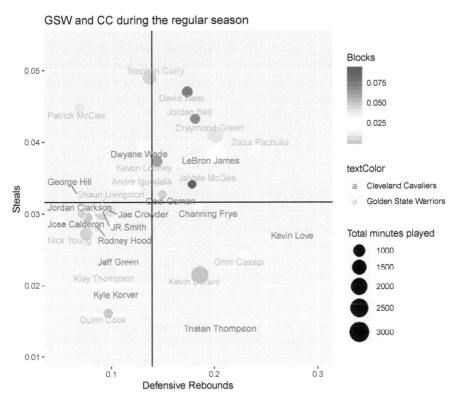

Figure 2.8   Bubble plot of the Golden State Warriors' and Cleveland Cavaliers' players according to statistics related to defense (per minute played).

Interesting evidence can be extracted from the graph of Figure 2.8. First, in the top-right quadrant (players with both defensive rebounds and steals above average), we predominantly find Golden State Warriors players having also excellent values for blocks. In their midst, the only noteworthy Cleveland Cavaliers player is LeBron James, with quite high defensive rebounds and steals, and blocks on average, though. It is worth mentioning David West and Jordan Bell, who exhibit excellent performances despite a low value of minutes played. On the other hand, in the bottom left quadrant (players with both defensive rebounds and steals below average) there are mainly Cleveland Cavaliers players who tend to have also low values for blocks. Among them we find one of the Golden State Warrior Splash Brothers, Klay Thompson. Kevin Durant exhibits defensive rebounds and blocks above average, but steals below average.

2.2.6 Variability analysis

In descriptive Statistics, variability is the extent to which data deviate from the average as well as the extent to which they tend to be different from each other. An index measuring variability is a nonnegative real number assuming value zero if all the data are the same and increases as the data become more diverse. For a given variable X, there are several indexes of variability, the most commonly used being the range (difference between maximum and minimum values), the interquartile difference (difference between third and first quartiles) and the variance σ_X^2 or its numerator, called deviance or Total Deviance (TD), used in Chapter 4. Very often, variability is also measured by the variance's square root, the standard deviation σ_X, already mentioned in Section 2.2.3, when introducing the standardization issue. In detail, σ_X^2 is obtained averaging the squared deviations of each of the N values x_i $(i = 1, \cdots, N)$ of variable X from their mean μ_X by means of the following formula:

$$\sigma_X^2 = \frac{\sum\limits_{i=1}^{N}(x_i - \mu_X)^2}{N},$$

and we then obtain the standard deviation as

$$\sigma_X = \sqrt{\sigma_X^2}.$$

When different variables have to be compared according to their variability, it is recommended to use a normalized index able to take account of the possibly different averages or unit of measurement of the compared variables. Typically, to this aim, we may resort to the variation coefficient (VC), given by the ratio of the standard deviation to the average ($VC = \sigma_X/\mu_X$). VC can be computed only when the values x_i are all positive[2] ($\min(x_i) > 0$).

In basketball, variability may concern the extent to which the players of a team perform differently from each other according to a given statistic. Variability may be a good thing or a bad thing. For example, with variables related to specialized tasks (*e.g.*, assists, rebounds, etc.), high variability may mean that the team is well balanced according to a given skill, in the sense that there are few players specifically devoted

---

[2]As a matter of fact, VC can also be computed with all negative values ($\max(x_i) < 0$), using the formula $VC = \sigma_X/|\mu_X|$. A variable with all negative values is quite uncommon in practice.

to that task, who are able to accomplish it much better than other players, devoted to other tasks. On the other hand, with variables related
to generic performance (*e.g.*, goal percentages), high variability means
that the team depends too much on a few players able to perform well,
while the others are far below the team standards.

Range, variation coefficient and standard deviation, together with
a useful variability diagram, can be obtained thanks to the function
`variability`, which allows to choose a number of variables whose variability has to be measured and represented. In the diagram, for each
variable, a vertical axis is drawn, where players are located at the level
corresponding to their value (on the selected variable) and are represented with a bubble with size proportional to another variable. The
extent to which the bubbles are scattered along the vertical axis gives
an immediate idea of that variable's variability. In addition, the values
assumed by the range and the variation coefficient are reported on the
graph for each variable.

Suppose we are interested in measuring the variability of the 3-point
shot percentages of the players of Oklahoma City Thunder who have
played at least 500 minutes. Table 2.5 shows the 3-point shots percentage
($P3p$) and 3-point shots attempted ($P3A$) for the selected players. With
the code

```
> Pbox.OKC <- subset(Pbox, Team=="Oklahoma City Thunder"
                   & MIN>=500)
> vrb1 <- variability(data=Pbox.OKC, data.var="P3p",
                   size.var="P3A")
```

we obtain the R object `vrb1` of class `variability`, a list containing
data frames of range, variation coefficient and standard deviation values, plus the data frames of the variable(s) whose variability has to be
measured (`data.var`) and the variables selected for drawing the bubble
size (`size.var`). The function has also the logical argument `VC`, that is
`TRUE` by default. It must be set to `FALSE` if the analyzed variable is not
strictly positive and the variation coefficient cannot be computed.

The average, standard deviation, variation coefficient and range of
variable $P3p$ result 30.41, 10.37, 0.34 and 40.07, respectively.

When studying variability, averages and standard deviations used to
compute the variation coefficient can be optionally weighted with a variable of interest. This can be done by adding the argument `weight=TRUE`
to the previous code; the weights will be given by the same variable

TABLE 2.5   3-point shots percentage
($P3p$) and 3-point shots attempted
($P3A$), Oklahoma City Thunder players.

Player	$P3p$	$P3A$
Russell Westbrook	29.75	326
Paul George	40.07	609
Carmelo Anthony	35.65	474
Steven Adams	0.00	2
Jerami Grant	29.09	110
Raymond Felton	35.22	230
Patrick Patterson	38.60	171
Alex Abrines	38.01	221
Andre Roberson	22.22	36
Josh Huestis	28.70	115
Terrance Ferguson	33.33	120
Corey Brewer	34.33	67

selected for the bubble size (`size.var`). In detail, the weighted mean
and the weighted standard deviation are computed as

$$_W\mu_X = \frac{\sum x_i \cdot w_i}{\sum w_i}$$

and

$$_W\sigma_X = \sqrt{_W\sigma_X^2} = \sqrt{\frac{\sum(x_i -_w \mu_X)^2 \cdot w_i}{\sum w_i}},$$

respectively, where w_i is the i-th weight, that is the i-th value of the variable in `size.var`. The idea is to weigh differently each observation (in the weighted mean) or each squared deviation from the average (in the weighted standard deviation), according to the values of the weighting variable. In our example, we select $P3A$ (see Table 2.5) as the weighting variable, and obtain weighted average, standard deviation and variation coefficient equal to 35.35, 4.30, 0.12, respectively. In this situation, since $P3p = P3M/P3A$, we note that the denominator equals the weight w_i, so that the weighted mean results to be the 3-point goal percentage of the whole team.

Extending the example, we focus again on the players of Oklahoma City Thunder who have played at least 500 minutes and measure variability of the variables denoting 2-, 3-point shots and free throw percentages, weighted by the corresponding number of attempted shots. With the following code

```
> vrb2 <- variability(data=Pbox.OKC,
                       data.var=c("P2p","P3p","FTp"),
                       size.var=c("P2A","P3A","FTA"),
                       weight=TRUE)
```

we first obtain the R object `vrb2` of class `variability`, analogous to the previous object `vrb1`. If only one variable is inserted in `size.var`, that variable is used to weigh all the variables in `data.var`. Alternatively, `size.var` can be filled in with the same number of variables inserted in `data.var` and the weighting variables weigh the analyzed variables according to their order. A plot method is available for this class, producing the above-described variability diagram. So, with the code

```
> plot(vrb, title="Variability diagram - OKC")
```

we obtain the graph of Figure 2.9 representing the variability diagram for the variables denoting 2-, 3-point shots and free throws percentages, with size of the bubbles proportional, respectively, to 2-, 3-point shots and free throws attempted.

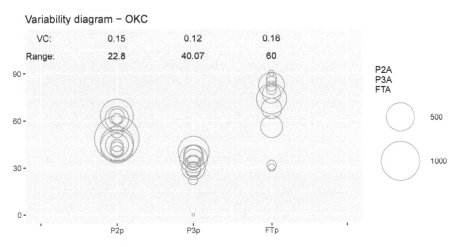

Figure 2.9 Variability diagram of the goal percentages $(P2p, P3p, FTp)$, weighted by the attempted shots $(P2A, P3A, FTA)$, Oklahoma City Thunder (OKC) players. VC = variation coefficient.

The most scattered bubbles are those related to free throws that exhibit the highest variability, as confirmed by range and variation coefficient. Here we even note a couple of outlier players. For 2-point shots, there are two groups of players with slightly different performance levels. Finally, 3-point shots exhibit the lowest variability, as confirmed by the variation coefficient. In this case, the range is not a reliable measure, as it is heavily affected by a single outlier player (Steven Adams) with only 2 shots attempted and 0 made[3].

In general, we note that the players with the best percentages are not always those attempting the highest number of shots.

2.2.7 Inequality analysis

Inequality is a concept originally born in economics, to represent the income or wealth distribution of a region or a country. In this context, inequality is the extent to which a small fraction of the population tends to own a big share of the total wealth. The extremes are represented by equal distribution (if everybody has the same wealth) or maximal inequality (when everybody has no wealth, except for just one person who owns the total wealth). We may borrow this concept from its native field and use it to investigate inequality within a basketball team, with reference to some game variable. In the following, we will refer to inequality of points made, but the same concept can be extended to any other variable. Specifically, inequality of points made occurs when there are few players who score a big part of the total points of the team (with maximal inequality meaning that all the points are scored by just one player). Conversely, equal distribution would imply that all the team members score the same points. High levels of inequality denote a team that is not well balanced from the point of view of the scored points and tends to depend too much on a few players.

In practice, both extreme situations are not reached. In order to measure and graphically represent how far a practical situation is from totally equal distribution, we use the Gini coefficient and the Lorenz curve. The Gini coefficient is an index ranging from 0 (equal distribution) to 100% (maximal inequality). Its mathematical definition is based on the Lorenz curve, which plots the fraction y of the total variable (the

---

[3]This examples gives us the opportunity to open a brief parenthesis about outliers, anomalous values often present in datasets. They can affect the analyses in different ways and usually need to be identified and treated. A good reference about this issue is Hawkins (1980).

TABLE 2.6   Example of inequality analysis: scored points by 8 players of the Oklahoma City Thunder.

Player	PTS	CPl	$CPTS$	$CPl\%$	$CPTS\%$
Patrick Patterson	318	1	318	12.50	3.98
Alex Abrines	353	2	671	25.00	8.39
Raymond Felton	565	3	1236	37.50	15.46
Jerami Grant	682	4	1918	50.00	23.98
Steven Adams	1056	5	2974	62.50	37.19
Carmelo Anthony	1261	6	4235	75.00	52.96
Paul George	1734	7	5969	87.50	74.64
Russell Westbrook	2028	8	7997	100.00	100.00

scored points in our context) on the y-axis that is cumulatively referred to the bottom fraction x of the population. Two lines are usually added to the graph, representing the two distributions with perfect equality and maximal inequality, respectively. A brief example can help understanding these two tools. Let us consider the first 8 players of the Oklahoma City Thunder, in decreasing order of scored points: from the smallest to the largest number of scored points (Table 2.6, Player and PTS). We first cumulate players and scored points (CPl and $CPTS$), then we divide respectively by the total number of players and the total scored points in order to obtain the cumulative percentages ($CPl\%$ and $CPTS\%$), informing us about the fraction of total points scored by the first fraction of players. For instance, we have that the first 25% players (2 over 8) scored 8.39% of the total points, the first 50% (4 players over 8) scored 23.98% of the total points, and so on.

The Lorenz curve is obtained by joining the points with coordinates given by $CPl\%$ (x-axis) and $CPTS\%$ (y-axis), as shown in Figure 2.10, where the perfect equality and maximal inequality lines are also plotted. The (blue shaded) area that lies between the line of equality and the Lorenz curve represents the so-called inequality area: the larger its size, the higher the inequality. The Gini coefficient can then be obtained as the ratio of the inequality area to the total area between the line of equality and the line of maximal inequality and is computed as

$$G = \frac{\sum_{i=1}^{N-1} (CPl\%_i - CPTS\%_i)}{\sum_{i=1}^{N-1} CPl\%_i} \cdot 100 = 38.12\%$$

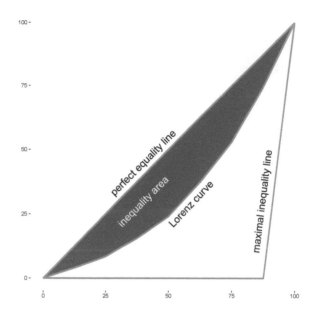

Figure 2.10 Example of inequality analysis: Lorenz curve.

where the subscript i is for the row number in Table 2.6 and N is the number of players considered in the analysis.

The function `inequality` allows us to represent the Lorenz curve and compute the Gini coefficient of a team, for a given number of players. Note that it makes sense to include in the analysis at most 8-9 players, in order to consider just the fundamental lineups. In fact, including players who don't play much inflates inequality without a specific interpretative advantage. For the same reason, comparisons among teams have to be made considering the same number of player for all the teams. If we consider 8 players, the teams with the lowest and highest values of the Gini coefficient are the Brooklyn Nets (9.59%) and Milwaukee Bucks (38.56%, almost tied with Oklahoma City Thunder), respectively. The results are shown in Figure 2.11, obtained using the following code:

```
> Pbox.BN <- subset(Pbox, Team=="Brooklyn Nets")
> ineqBN <- inequality(Pbox.BN$PTS, nplayers=8)
> Pbox.MB <- subset(Pbox, Team=="Milwaukee Bucks")
> ineqMB <- inequality(Pbox.MB$PTS, nplayers=8)
> library(gridExtra)
> p1 <- plot(ineqBN, title="Brooklyn Nets")
```

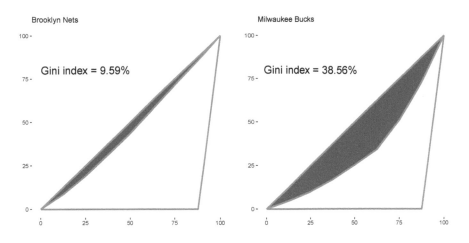

Figure 2.11 Inequality analysis of scored point of Brooklyn Nets and Milwaukee Bucks - 8 players.

```
> p2 <- plot(ineqMB, title="Milwaukee Bucks")
> grid.arrange(p1, p2, nrow=1)
```

Note that `ineqBN` and `ineqMB` are R objects of class `inequality` containing the Gini index value and the cumulative distributions used to plot the Lorenz diagram. A plot method is available for this class, producing the Lorenz curve with the annotated value of the Gini index.

An interesting further analysis can be carried out by jointly considering the Gini coefficients and the total points made by the teams. We write a `for` cycle in order to compute the Gini coefficients for all the teams:

```
> no.teams <- nrow(Tbox)
> INEQ <- array(0, no.teams)
> for (k in 1:no.teams) {
        Teamk <- Tbox$Team[k]
        Pbox.sel <- subset(Pbox, Team==Teamk)
        index <- inequality(Pbox.sel$PTS, npl=8)
        INEQ[k] <- index$Gini
        }
```

Then we obtain the scatter plot of the Gini coefficient (x-axis) against the points made (y-axis) using the `scatterplot` function presented in Section 2.2.4:

```
> dts <- data.frame(INEQ, PTS=Tbox$PTS,
                    CONF=Tadd$Conference)
> mypal <- colorRampPalette(c("blue","red"))
> scatterplot(dts, data.var=c("INEQ","PTS"), z.var="CONF",
              labels=Tbox$Team, palette=mypal,
              repel_labels=TRUE)
```

The obtained graph - where the teams have been colored according to
the Conference - shows that, in general, teams with higher inequality
indexes tend to score more points (Figure 2.12). Of course this has to
be interpreted just as a general tendency, not as a rule. Indeed, we have

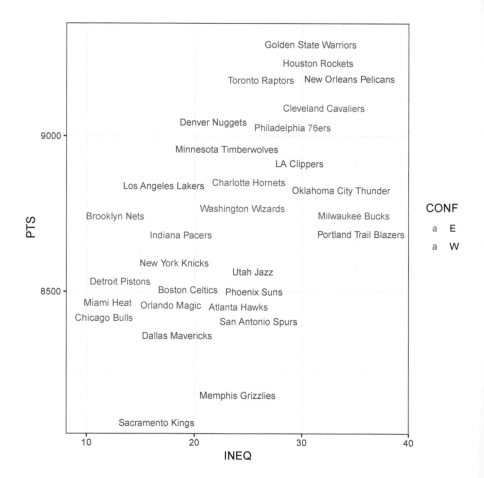

Figure 2.12 Gini coefficient (INEQ, x-axis) against total points made
(PTS, y-axis) in the Western (red) and Eastern (blue) Conference.

a counterexample: the two teams with the lowest and the highest Gini index (Brooklyn Nets and Milwaukee Bucks, respectively) have scored almost the same number of points.

Another interesting application of the inequality analysis is with reference to the points scored by a given lineup on the field. For example, we may be interested in the lineup composed by Stephen Curry, Kevin Durant, Klay Thompson, Draymond Green and Zaza Pachulia. In order to compute the points scored by these players when that particular lineup was on the field, we need to resort to play-by-play data. We select, from the play-by-play dataset PbP (obtained by manipulating PbP.BDB as described on page 23) only the plays by the Golden State Warriors

```
> PbP.GSW <- subset(PbP, team=="GSW")
```

then we identify the rows referring to events occurred when the above mentioned lineup was on the field

```
> lineup <- c("Stephen Curry", "Kevin Durant",
              "Klay Thompson", "Draymond Green",
              "Zaza Pachulia")
> filt5 <- apply(PbP.GSW[, 4:13], 1,
              function(x) {
              x <- as.character(x)
              sum(x %in% lineup)==5
              })
```

we count the points scored by the five players in those moments

```
> subPbP.GSW <- PbP.GSW[filt5, ]
> PTS5 <- sapply(lineup,
              function(x) {
              filt <- subPbP.GSW$player==x
              sum(subPbP.GSW$points[filt], na.rm=T)
              })
```

and we finally perform the inequality analysis

```
> inequality(PTS5, nplayers=5)
```

The Gini coefficient results 16.97%, denoting an appreciable tendency to equality in the points scored by these five players, when exactly this lineup is on the field. Of course, this may change from a match to another. The analysis can be repeated for single matches: we just need to select the rows referred to a given opponent and carry out the same analysis. For example, considering the match against Detroit Pistons, we use the dataset

```
> PbP.GSW.DET <- subset(PbP, team=="GSW" & oppTeam=="DET")
```

and we obtain a Gini coefficient of 48.44%, with Klay Thompson accounting for the main part of the points scored by the lineup.

A different approach to study whether the points scored (or another game variable) is concentrated in the hands of a few players is by computing heterogeneity or diversity indices. In this situation, the different players are considered as the categories among which the total amount of scored points are distributed. Each player is responsible for a proportion (relative frequency) of the scored points and heterogeneity indices measure how much different are such relative frequencies. Examples of heterogeneity indices are the Gini index (also known as Gini-Simpson index, that is different from the inequality Gini index) and the Shannon (also known as Shannon-Wiener) index. Both of them, in their normalized version, equal 0 (minimum heterogeneity) when one single player scored all the points (one single category has relative frequency equal to 1) and equal 1 (maximum heterogeneity) when the scored points are equally distributed among the players (each category has the same relative frequency).

2.2.8 Shot charts

Shot charts are a useful tool to show shot patterns across the court, in order to analyze the players' favorite spots on the court to shoot from, also with reference to selected game phases. The play-by-play dataset PbP, obtained from PbP.BDB thanks to the function PbPmanipulation as described on page 23, contains different types of shot coordinates (see Table 2.2) that can be graphically analyzed with shot charts.

The R function shotchart creates a customizable plot where each shot is displayed as a point on the court, possibly colored according to whether the shot is made or missed. Additionally, the court can be split into a given number of sectors that can be colored according to some aggregate statistics (*e.g.*, average scored points, average play length when the shot is attempted, ...) and annotated with shots made/attempted and field goal percentages.

The function shotchart is built with reference to the standard 94 by 50 feet NBA court. It requires as input half-court shot coordinates expressed in feet, with the origin located in the center of the court. This format can be obtained by a linear transformation of original coordinates contained in variables `original_x` and `original_x` of PbP measured by the tracking systems. In detail, suppose we are interested to analyze Kevin Durant's shot patterns. We preliminarily select and transform data as follows:

```
> subdata <- subset(PbP, player=="Kevin Durant")
> subdata$xx <- subdata$original_x/10
> subdata$yy <- subdata$original_y/10-41.75
```

where 41.75 is the distance of the hoop center from the center of the court, as `original_x` and `original_y` have (0,0) in the center of the basket (see Table 2.2). The plot representing only the scatter of shots (Figure 2.13) can be obtained by

```
> shotchart(data=subdata, x="xx", y="yy", type=NULL,
            scatter=TRUE)
```

Both the points and the background colors can be customized by means of the arguments `pt.col` and `bg.col`, respectively. Points can be colored differently according to whether they are made or missed by means of

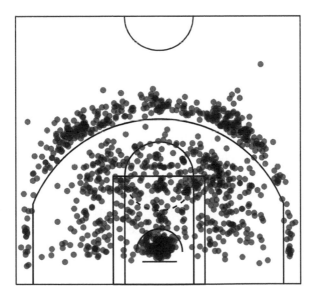

Figure 2.13 Shot chart (Kevin Durant).

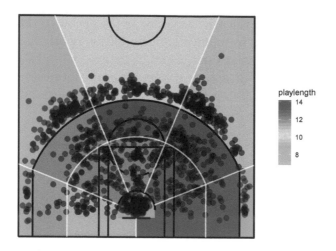

Figure 2.14 Shot chart with 16 areas colored according to the length of the play (Kevin Durant).

the option z="result", where z is the argument of shotchart allowing to specify a third variable to analyze, and result is the variable of PbP containing the information about the result of the shot (an example using this feature will be shown later, in Chapter 4, Section 4.2.2, Figure 4.6). With the option type="sectors", the court is split into num.sect slices, that can be colored according to the value assumed by a numerical variable specified by the argument z. In detail, with the code

```
> shotchart(data=subdata, x="xx", y="yy", z="playlength",
            num.sect=5, type="sectors", scatter=TRUE)
```

we obtain the chart of Figure 2.14, highlighting, for each zone, the average time elapsed since the immediately preceding event when the shot is made. From this chart, we note that Kevin Durant's close-range shots occur on average in the first 8-9 seconds of the play, while as the play length increases, he more commonly attempts mid-range shots from the center or his left-hand side. By setting scatter=FALSE and result="result", the scatter is replaced by the annotation of shots made/attempted and field goal percentages (Figure 2.15). In this graph, we note that close-range shots are the most successful ones and that the zones where late shots are attempted tend to have lower percentages. The option type also allows to draw plots of shot density across the court, as will be discussed in Chapter 3, Section 3.5.2.

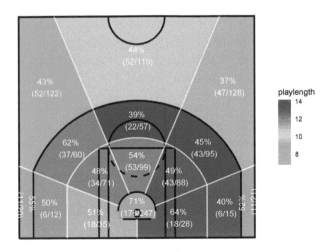

Figure 2.15 Shot chart with 16 areas colored according to the length of the play and annotated with shooting statistics (Kevin Durant).

GLOSSARY

Box scores: A box score is a summary of the results from a sport competition, structured according to a set of game variables that are different depending on the sport. The box score data are usually derived from a statistics sheet recorded manually or with the help of proper equipments, and are then summarized into frequency tables or averages, referring to the team or the single players' achievements.

Data warehouse: Computer archive containing data, designed to allow easy production of analyses and reports useful for decision-making purposes. Data are typically cleansed, transformed, listed and made available for data mining, analytical processing and decision support.

Distributed system: A system whose components are distributed on different networked computers, which share resources and interact with each other in order to achieve a common goal.

Inequality: In Statistics, inequality is a feature of the distribution of numerical transferable (*i.e.*, whose units can conceptually be moved from one subject to another) variables and is related to the extent to which a big share of the analyzed variable tends to be concentrated in a small fraction of the cases. In basketball, inequality analysis can be used to measure the dependence of the team on a small number of players from the point of view of the scored points or some other game variable. Traditionally, inequality is analyzed by means of the Gini coefficient and the Lorenz curve.

Play-by-play data: A play-by-play is a set of data that recounts each play of a game as it occurs or occurred, along with relevant information about the recorded event (*e.g.*, time, player(s) involved, score, area of the court, etc.).

Relational database: Digital database where data are organized into one or more tables of columns and rows, with a unique key identifying each row.

Scalable system: A system, network or process that is capable to manage a growing amount of work, or has potential to be enlarged to handle that growth.

Variability: In Statistics, variability refers to how dispersed or spread out the data values are. Variability is absent if all values are the same, and it increases as long as they are more and more different from each other, or fluctuating around a measure of central tendency. There are several indexes of variability, the most commonly used being the (interquartile) range, the variance (or its square root, the standard deviation, or its numerator, the deviance or Deviance TD) and the variation coefficient (ratio of the standard deviation to the average).

Web scraping, web harvesting, web data extraction: Gathering and extraction of data from websites into a central local database or spreadsheet, for later retrieval or analysis, both manually and with automated informatics processes. Web scraping may be against the terms of use of some websites.

II

Advanced Methods

CHAPTER 3

Discovering Patterns in Data

A PATTERN is an intelligible regularity discernible in the way in which something happens. To discover patterns in data is about using Data Science to reveal the hidden mechanisms governing the analyzed phenomena. Data Science is able to uncover lots of different patterns such as distributions, associations, similarities, interactions, classifications and trends, that can be described, measured and modelled.

> There are only patterns, patterns on top of patterns, patterns that affect other patterns. Patterns hidden by patterns. Patterns within patterns. If you watch close, history does nothing but repeat itself.
>
> Chuck Palahniuk
> American novelist and freelance journalist

In this chapter, we will discuss some examples of patterns that can be revealed and many statistical tools that can be exploited to analyze data. The focus topic in the end of this chapter will be concerned with the identification of the factors determining high-pressure game situations and affecting the scoring probability. Two important issues in pattern discovery, namely finding groups in data and modeling relationships in data deserve to be treated in two separate specific chapters (Chapters 4 and 5).

3.1 QUANTIFYING ASSOCIATIONS BETWEEN VARIABLES

With the term dependence or association we usually mean any statistical relationship, whether causal or not, between two or more variables (Liebetrau, 1983; Merlo and Lynch, 2010). In order to detect any association, all the variables must be jointly measured on the units of observations. Leaving aside the concept of causality, which would require a specific discussion, here we refer to the general notion of association that can be defined in several different ways, according to which conception of relationship is assumed. In basketball, discovering associations within a set of variables can be interesting from different points of view. For example, we may reveal which game variables are most associated with the scored points, in general or with reference to a single team or player or we may even detect associations between two specific game variables that tend, for example, to enhance or undermine each other.

In most statistics textbooks, methods for detecting association among variables are usually classified according to the nature of the variables involved. We can distinguish categorical (or qualitative) and numerical (or quantitative) variables. Categorical variables assign each unit of observation to a particular group or category on the basis of some qualitative property; possible values for categorical variables are then attributes, categories. Examples of categorical variables are the Conference to which a team belongs (East or West), the result of a shot (made or missed) or the player's role (with five possible values given by point guard, shooting guard, small forward, power forward, and center). On the contrary, possible levels for numerical variables are numbers (integers, in the case of discrete numerical variables, or intervals, in the case of continuous numerical variables). The total number of points made is an example of discrete numerical variable, while the time played in the quarter is a continuous numerical variable.

The distinction between categorical and numerical variables is important because statistical methods conceived for variables of one kind cannot be appropriate to analyze the other kind of variables. Actually, there exists a hierarchy, according to which methods for categorical variables can be used to analyze numerical variables (the latter having a higher information content), and not vice versa. Despite this, due to the wide range of methods available, it is always a better choice to resort to the statistical methods especially introduced in the literature for treating with the type of variable at hand.

Simultaneous examination of relationships among several variables can be addressed by several methods of multivariate data analysis (Härdle and Simar, 2015). Focusing now on the bivariate analysis, the three most common definitions of association between two variables are statistical dependence (Section 3.1.1), mean dependence (Section 3.1.2) and correlation (Section 3.1.3), which can be evaluated by means of proper statistical indexes and tests. According to the nature of the analyzed variables, statistical dependence can be evaluated when variables are both categorical, or at least one is numerical, or they are both numerical. Studying the mean dependence requires at least one numerical variable. Correlation analysis requires that both variables are numerical. In this book, more space will be devoted to the study of correlation, because in basketball analytics the variables of interest are often numerical.

3.1.1 Statistical dependence

Statistical dependence can be studied starting from a two-way cross-table and investigating the existence of a "general" relationship among the two variables (see, among others, Fagerland et al., 2017). In particular, this method compares the observed number of units within the cross-table cells (observed frequencies) with the number of units one would expect in the cells if no association exists and the units were randomly distributed (expected frequencies under the independence hypothesis). Several association indexes proposed in the literature are based on a summary of the differences between the observed and the expected frequencies: there is association between the two variables when the observed and expected frequencies differ statistically (beyond random chance). Different approaches to measure association can be set up, for example by measuring the reduction in the error of prediction for a variable thanks to the knowledge of a second variable.

The most famous association index based on the differences between observed and expected frequencies is the Chi-square (X^2, sometimes indicated as χ^2). Consider a table with r rows and c columns and let n_{ij} and $\hat{n}_{ij}$ ($i = 1, \ldots, r$; $j = 1, \ldots, c$) the observed and expected frequency in the generic cell i, j, respectively. X^2 can be computed as $\sum_i \sum_j \frac{(n_{ij} - \hat{n}_{ij})^2}{\hat{n}_{ij}}$. Since it measures the intensity of the relationship between the two variables, but also depends on the sample size N and the number r of rows and columns c in the table, this statistic has been adjusted in several ways, giving birth to a number of related measures of association (see, among others, Warner, 2013), like Phi or Φ ($\Phi = \sqrt{X^2/N}$, also known as $M_2(D)$), the Mean Squared Contingency (Φ^2), Pearson's contingency coefficient P ($P = \sqrt{\Phi^2/(\Phi^2+1)}$), Cramer's V, also known as the normalized index C ($V = \Phi/\sqrt{k-1}$, where $k = \min(r, c)$). When there is no association between the two variables, each of these measures has a value of 0. As the intensity of association increases, the value of each of these measures increases. Cramer's V is the preferred measure because it is the only one that equals 1 in the case of a perfect association between the two variables and so can be easily interpreted as a percentage. In addition, since Cramer's V formula considers the dimensions of the table, it can be used for comparisons among tables of different dimensions. A significance test is usually performed on X^2 (called Chi-square test of independence, based on the χ^2 distribution) in order to test whether the X^2 value can be considered statistically different from zero, indicating that there exists a significant association between the two variables. For the other association measures derived from X^2, it can be said that their tests of significance lead to the same conclusion as it is for the chi square test of independence.

For example, we may be interested in analyzing if some game statistics of the Golden State Warriors depends on the opponent team. To do this, we can compute some association measures between the two variables crossed in Table 3.1, which reports the number of free throws, missed shots, rebounds and attempted shots by the Golden State Warriors in the matches played against each of the opponent teams shown in the table rows.

Table 3.1 is obtained by selecting, from the play-by-play dataset PbP (obtained by manipulating PbP.BDB as described on page 23) only the plays by the Golden State Warriors

```
> PbP.GSW <- subset(PbP, team=="GSW")
```

and then by filtering out some game events into which we are not interested

```
>ev <- c("ejection","end of period","jump ball",
         "start of period","unknown","violation",
         "timeout","sub","foul","turnover")
> event.unsel <- which(PbP.GSW$event_type %in% ev)
> PbP.GSW.ev <- PbP.GSW[-event.unsel,]
```

TABLE 3.1   Opponent teams (rows) and event types (columns), Golden State Warriors play data.

	free throw	miss	rebound	shot
ATL	33	88	81	84
BKN	34	80	98	93
BOS	45	95	90	71
CHA	26	91	90	80
CHI	46	80	98	95
CLE	47	88	95	79
DAL	74	155	188	188
DEN	78	172	164	173
DET	34	75	85	83
HOU	56	118	119	131
IND	33	97	90	72
LAC	127	161	166	176
LAL	104	190	202	176
MEM	77	126	128	117
MIA	48	92	92	79
MIL	33	70	74	85
MIN	54	132	142	133
NOP	85	183	175	180
NYK	46	78	79	90
OKC	86	176	179	153
ORL	27	77	92	99
PHI	46	76	88	98
PHX	59	166	178	197
POR	60	123	116	125
SAC	76	165	169	159
SAS	70	175	169	162
TOR	50	68	66	90
UTA	64	190	167	158
WAS	50	83	89	83

The cross-table T (reported in Table 3.1) is given by

```
> attach(PbP.GSW.ev)
> T <- table(oppTeam, event_type, exclude=ev)
> detach(PbP.GSW.ev)
```

and some association measures can be directly obtained by resorting to the function assocstats in the library vcd (Meyer et al., 2017), as follows:

```
> library(vcd)
> assocstats(T)
```

Results show that the association between the two variables is low (Cramer's V equals 0.06) but significantly different from zero (Pearson's $X^2 = 116.25$ has a p-value equal to 0.011: association is usually considered significant when p-value is lower than conventional values 0.05 or 0.10). The other results provided by the function `assocstats` are the Likelihood ratio, also known as G-test, that gives here the same indications as the Chi-square test of independence (details can be found in Hosmer et al., 2013), the Phi coefficient (available in this function only for 2×2 tables) and Pearson's contingency coefficient ($P = 0.097$). We may conclude that there is a (low) association and the number of game events (shots, missed shots, rebounds and free throws) in the play of the Golden State Warriors depends on the opponent team (interpretation should pay attention to the fact that teams do not play against all other teams an equal number of times).

Another example of computation of Cramer's V is reported in Chapter 3, Section 3.4 and concerns the relationship between assists made and received (Table 3.4).

3.1.2 Mean dependence

The mean dependence method allows us to examine, for example, if the average number of points scored by all the NBA teams differ between the East and West Conferences or among the different Divisions, or between qualified and non-qualified teams for Playoffs, or to study if the average number of fouls (or assists, rebounds, ...) of one team differs across the quarters. In those situations, we want to analyze if and how the average of a numerical variable (*e.g.*, the points made) varies across the classes defined by another variable, which can be categorical (*e.g.*, Playoff, with values Yes or No).

In the field of exploratory data analysis[1], a variable Y is said to be mean independent from another variable X if and only if the conditional means of Y (that is, the means computed within each group or class defined by X) are all equal and, consequently, equal to the unconditional mean of Y (computed over all the observation units, without considering their classification according to X). A well-known index able to measure the level of mean dependence of Y with respect to X is the Pearson's correlation ratio $\eta^2_{Y|X}$, that is the ratio of the between

---

[1]In inferential statistics, unlike descriptive or exploratory statistics, mean values obtained from different groups are usually compared by means of t-tests and ANOVA tables (see, among others, Hand and Taylor, 1987).

deviance over the total deviance (BD/TD). This formulation is based on the Total Deviance decomposition that will be explained in Chapter 4, Section 4.1. It ranges from 0 (when the conditional means of Y are all equal and Y is mean independent from X) to 1 (Y perfectly depends, on average, on X: in other words, for each observation, to know the class or group defined by X to which the observation belongs is enough to know also the value of Y). Pearson's correlation ratio finds a very useful application in cluster analysis, where it helps deciding how many clusters to maintain in the solution of a k-means clustering, as explained in Chapter 4, Sections 4.1 and 4.2.1.

In order to investigate the mean dependence of some game variables on Playoff qualification, we computed the conditional means of each game variable, that is averaging over teams qualified (Playoff=Yes) and not qualified (Playoff=No), separately, and the values of the Pearson's correlation ratio η^2, in %, as displayed in Table 3.2. The game variables considered are points made, 2-point and 3-point field goals made, free throws made, total rebounds, assists, steals, blocks, Defensive and Offensive Ratings. The values in Table 3.2 are obtained (in the object `eta`) running the following code, which depends on the libraries `dplyr` (Wickham et al., 2019) and `lsr` (Navarro, 2015) and makes use of the pipe operator `%>%` in the library `magrittr` (Bache and Wickham, 2014):

TABLE 3.2 Game variables' averages (conditional to Playoff) and values of η^2 of mean dependence of game variables on Playoff qualification.

| | Playoff | | |
	No	Yes	$\eta^2\%$
DRtg	107.90	104.60	42.53
ORtg	104.00	108.10	40.25
STL	601.90	659.60	28.77
PTS	8576.00	8844.80	19.28
BLK	365.60	420.40	18.12
FTM	1328.00	1394.40	5.58
P2M	2353.70	2417.20	3.28
AST	1875.50	1931.60	3.17
P3M	846.90	871.90	1.07
REB	3558.10	3577.50	0.49

```
> library(dplyr)
> library(lsr)
> library(tibble)
> FF <- fourfactors(Tbox, Obox)
> attach(Tbox)
> attach(FF)
> X <- data.frame(PTS, P2M, P3M, FTM, REB=OREB+DREB, AST,
                  STL, BLK, ORtg, DRtg)
> detach(Tbox)
> detach(FF)
> Playoff <- Tadd$Playoff
> eta <- sapply(X, function(Y){
    cm <- round(tapply(Y, Playoff, mean), 1)
    eta2 <- etaSquared(aov(Y~Playoff))[1]*100
    c(cm, round(eta2, 2))
  }) %>%
>   t() %>%
>   as.data.frame() %>%
>   rename(No=N, Yes=Y, eta2=V3) %>%
>   rownames_to_column('rownm') %>%
>   arrange(-eta2)
>   column_to_rownames('rownm')
```

All the conditional means differ between qualified and non-qualified teams. Therefore, we can conclude that all these game variables are, to some extent, dependent on the Playoff qualification. However, as shown by the values of η^2, in several cases the degree of the mean dependence is very low. For example, the number of rebounds and assists or the number of shots made (2-point, 3-point and free throws) are substantially not dependent on the Playoff qualification (all η^2 lower than 6%). On the contrary, Defensive and Offensive Ratings are (moderately) highly dependent on Playoff ($\eta^2 = 42.53\%$ and 40.25%, respectively). These results show that it is the game as a whole that counts, not the game variables taken individually. Single variables don't tell the whole story; for example, the points made are only one side of the coin, the points taken must also be considered.

3.1.3 Correlation

Correlation is a specific kind of statistical association which refers to the linear relationship between two numerical variables. When numerical variables are available, measuring the degree of association by means of statistical dependence or mean dependence methods means to degrade both variables (in the case of statistical dependence) or one of them (in the case of mean dependence) to fill the role of categorical variables. Instead, correlation analysis allows the optimal use of the information available in the numerical variables, which, as already mentioned, in basketball analytics, and especially in performance analysis, prevail over categorical ones.

In much detail, correlation analysis is based on concordance indices that are positive when the highest (lowest) values of one variable are associated with the highest (lowest) values of the other variable, and negative when, on the contrary, the highest (lowest) values of one variable are associated with the lowest (highest) values of the other variable.

The most widespread concordance index is Pearson's correlation coefficient, which is designed to measure both intensity and direction of a linear relationship between the two variables. The focus is then on a linear relationship, and we can think of an interdependence between variables. Other concordance indexes measuring nonlinear association between variables have been proposed, for example the well-known Kendall's τ.

Given two variables X and Y, the value of Pearson's correlation coefficient ρ between X and Y ranges from -1 to 1, with the extremes meaning perfect (negative or positive, respectively) correlation and values near to 0 denoting absence of linear correlation (but not necessarily absence of any kind of association). Figure 3.1 displays some possible configurations of scatterplots for some values of ρ. It is clear that the higher the absolute value of ρ, the higher the intensity of (positive or negative, according to the sign) linear relationship. It must be said that there are many other possible configurations associated with a null correlation, which means absence of a linear relationship between X and Y and this is consistent with the existence of a nonlinear relationship between the two variables (Matejka and Fitzmaurice, 2017).

The following code computes the value of ρ between the number of assists, AST, and turnovers, TOV, (per minute played) for players who have played at least 500 minutes:

```
> data <- subset(Pbox, MIN>=500)
> attach(data)
> X <- data.frame(AST, TOV)/MIN
> detach(data)
> cor(X$AST, X$TOV)
```

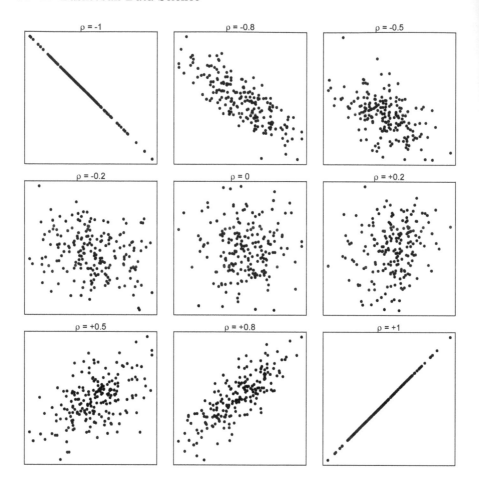

Figure 3.1 Configurations of scatterplot with different values of ρ.

The Pearson correlation coefficient equals 0.687: as expected, there exists a positive linear relationship between assists and turnovers and the intensity of the association is rather strong.

A very interesting interpretation of Pearson's correlation coefficient is given when it is used to measure the association between the rankings of players based on the two analyzed variables. For example, we first consider the ranking of players based on assists made (per minute played). The top assistmen are Rajon Rondo (New Orleans Pelicans), Russell Westbrook (Oklahoma City Thunder), John Wall (Washington Wizards), while the bottom ranks are occupied by Dante Cunningham (New Orleans Pelicans), Josh Huestis (Oklahoma City Thunder), and lastly, Semi Ojeleye (Boston Celtics). Another ranking can be obtained considering the turnovers (per minute played): here, the top players are DeMarcus Cousins (New Orleans Pelicans), Russell Westbrook (Oklahoma City Thunder) and James Harden (Houston Rockets), while on the bottom we find Terrance Ferguson (Oklahoma City Thunder), Dante Cunningham (New Orleans Pelicans) and Rodney Hood (Cleveland Cavaliers).

In order to measure the agreement between the two rankings, we can proceed with the previous example and compute the Pearson's correlation coefficients between rankings, as follows:

```
> cor(rank(X$AST), rank(X$TOV))
```

This is equivalent to compute Spearman's correlation coefficient between the two variables AST and TOV (on which rankings are based). Spearman's correlation coefficient is one of the most common rank-correlation measures and ranges from -1 to 1: it equals 1 when the players' positions are identical in the two rankings (perfect rank-agreement) and -1 when one ranking is the reverse of the other (perfect disagreement). Values close to 0 suggest no association between rankings and increasing values imply increasing agreement between rankings.

In our example, we have

```
> cor(X$AST, X$TOV, method="spearman")
```

The rank-correlation measure results 0.668, denoting a positive and strong association between the two rankings: players on top positions in the ranking of assists tend to rank high also on turnovers.

If we have more than two variables, we can compute all the Pearson correlation coefficients between each pair of variables and collect them in a matrix, called correlation matrix. It is a squared matrix (number of rows equal to the number of columns), with dimension given by the number of analyzed variables. Its generic i, j element is the Pearson correlation coefficient between the i-th and the j-th variable, ρ_{ij}. Each element on the principal diagonal of the correlation matrix is the correlation of a variable with itself, which is always 1. The off-diagonal elements vary between -1 and 1 and, since $\rho_{ij} = \rho_{ji}$, the correlation matrix is said to be symmetric and usually only the upper (or lower) triangle is shown.

If we give a data frame as input to the `cor` function (instead of two single variables, as shown before), we get a correlation matrix. For example, in the previous example

```
> cor(X)
```

returns the following 2×2 correlation matrix

	AST	TOV
AST	1.000	0.697
TOV	0.697	1.000

Obviously, if X contains more than two variables, cor returns a correlation matrix of higher dimension. Nice representations of the correlation matrix are available in the R Package BasketballAnalyzeR, as shown in Section 3.2, where the function corranalysis is introduced.

3.2 ANALYZING PAIRWISE LINEAR CORRELATION AMONG VARIABLES

A study of the linear correlation between all the pairs of variables in a given set can be performed thanks to the functions corranalysis and scatterplot, the latter already introduced in Chapter 2, Section 2.2.4. The term "pairwise" indicates that, even if more than two variables are jointly analysed with these two functions, leading to the creation of a correlation matrix and its related plot, the linear correlation coefficient remains a measure of bivariate association, as it evaluates direction and intensity of the linear relationship between all the possible pairs of variables.

After merging the datasets Pbox and Tadd in order to complete the players' boxscores data with some additional variables regarding the teams (conference, division, qualification for Playoffs, ...), we select those players who have played at least 500 minutes.

```
> data <- merge(Pbox, Tadd, by="Team")
> data <- subset(data, MIN>=500)
```

By way of example, we consider the following variables: scored point, 3- and 2-point shots made, total rebounds (offensive and defensive), assists, turnovers, steals and blocks (per minute played).

```
> attach(data)
> X <- data.frame(PTS, P3M, P2M, REB=(OREB+DREB), AST,
                  TOV, STL, BLK)/MIN
> X <- data.frame(X, Playoff=Playoff)
> detach(data)
```

We first compute pairwise linear correlation between the numerical variables and plot a graphical representation of the correlation matrix thanks to the function `corranalysis` and the corresponding `plot` method[2].

```
> corrmatrix <- corranalysis(X[,1:8], threshold=0.5)
> plot(corrmatrix)
```

The top panel of Figure 3.2 gives a graphical description of the correlation matrix, a square grid whose i, j entry is the Pearson correlation coefficient between the i-th and the j-th variable of the list, ρ_{ij}. In the lower triangle of the grid, we can read the values of the Pearson correlation coefficients, whereas the upper triangle uses stylized ellipses to give a rough idea of the shape of the relationship corresponding to each coefficient. A colored blue-red scale is also used and displayed on the right of the grid, where blue and red denote respectively negative and positive correlation, with intensity proportional to the correlation strength. The function `corranalysis` automatically performs a statistical test on the null hypothesis $\rho_{ij} = 0$, at the significance level $1 - \alpha = 95\%$, and all the coefficients that turn out to be not significantly different from zero are marked with a cross on the grid. In simple words, a cross means that the correlation coefficient between the two corresponding variables cannot be considered statistically different from zero. Instead, every value displayed on the grid is statistically different from zero, even if it is very close to zero. The significance level of the test can be tuned using the argument `sig.level`.

The bottom panel of Figure 3.2 shows the correlation network, where the pairs of variables whose Pearson's correlation coefficient is significant are joined by a blue or red (according to whether the correlation is negative or positive) edge with color intensity proportional to the absolute value of the correlation coefficient. The function also allows to set the argument `threshold` that deletes from the correlation network all the edges corresponding to Pearson's correlation coefficients with absolute value lower than a given threshold. In the examined example, we display only correlations higher than 0.5, positive or negative (`threshold=0.5`).

We detect, as expected, a positive correlation between blocks and rebounds, with a negative correlation between 3-point shots made

---

[2]Note that Figure 3.2 is obtained by setting the argument `horizontal=FALSE` in the `plot` command. If `horizontal=TRUE` (default option), the two plots appear side by side.

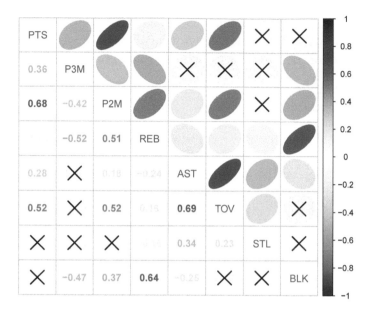

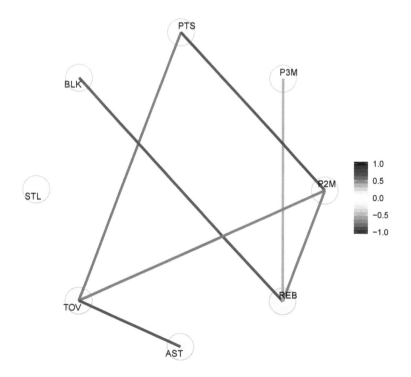

Figure 3.2 Linear correlation matrix plot.

and rebounds. Another well-known evidence is the positive correlation between assists and turnovers. More subtle remarks concern the positive correlation between turnovers and scored points, and the low correlation between 3-point shots made and scored points, which instead turn out to be much more correlated to 2-point shots. According to these evidences, we may conjecture that players able to move the ball (even if it entails the side effect of a higher number of turnovers) and to shoot from within the 3-point line have been very effective in terms of scored points in the 2017/2018 NBA championship.

Another tool for a graphical correlation analysis is the function scatterplot that, when three or more variables are given an input, produces a scatter plot matrix showing in the upper triangle the values of the Pearson correlation coefficients. With respect to the graph in the top panel of Figure 3.2, the main differences are that (1) the scatter plots are displayed in place of the stylized ellipses, (2) a categorical grouping variable can be added in order to analyze within-group correlations and (3) no significance analysis is performed. In our example, we may be interested to separate the players in two groups according to whether their team qualified for the Playoffs or not.

```
> scatterplot(X, data.var=1:8, z.var="Playoff",
              diag=list(continuous="blankDiag"))
```

The argument diag is specified so as to produce an empty diagonal, but other choices are possible, as will be described later, together with other options available for the arguments upper and lower, allowing to customize the upper and lower triangle of the scatter plot matrix (see Chapter 3, Section 3.5 and Chapter 5, Section 5.2). In the graphs of Figure 3.3 each point is a player, color-coded according to the qualification for Playoff (Yes or No, Y/N) of his team. The lower triangle shows the scatter plots of all the pairs of variables, the upper triangle contains the Pearson correlation coefficients, separately for the two groups. It is interesting to note that when a pair of variables has positive correlation, this is almost always higher for players in teams that qualified for Playoffs and the opposite happens for negative correlations. It seems that players in the best performing teams tend to strengthen the positive synergies between their game variables and soften the negative ones. The last two rows and the last column in Figure 3.3 display univariate analyses of the game variables: respectively, the histograms (separately for the two groups) and the boxplots. Through these univariate analyses, we do not detect relevant differences in game variables between the two groups.

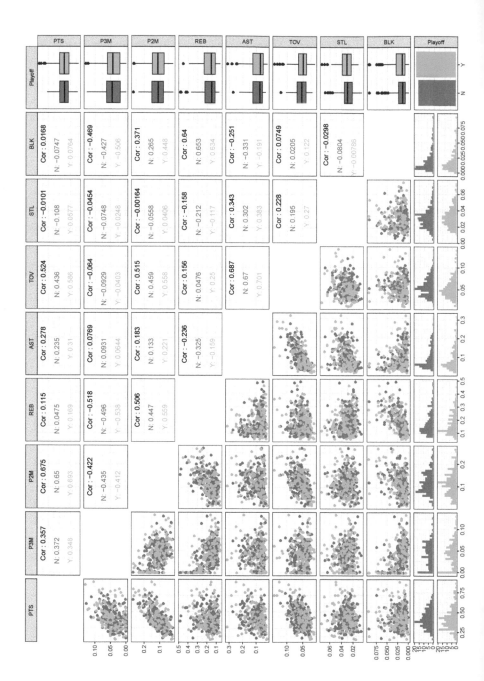

Figure 3.3 Scatter plot matrix with within-group analysis.

3.3 VISUALIZING SIMILARITIES AMONG INDIVIDUALS

There exist several methods of multivariate data analysis that allow to visualize similarities among individuals, for example (multiple) Correspondence Analysis (Greenacre, 2017), Principal Components Analysis (Jolliffe, 1986), optimal scaling methods (Gifi, 1990; de Leeuw and Mair, 2009). Here we focus on Multidimensional Scaling (MDS), which is a nonlinear dimensionality reduction tool that allows to plot a map visualizing the level of similarity of individual cases of a dataset. In detail, the starting point is a dissimilarity matrix $\Delta = (\delta_{ij})_{i,j=1,...,N}$, where δ_{ij} represents the dissimilarity between case i and case j. Dissimilarities are distance-like quantities that can be obtained in several ways. When a distance metric is used, Δ is a distance matrix and is usually denoted by $\mathbf{D}^p = (d_{ij})_{i,j=1,...,N}$ in order to distinguish it from the more general case. The superscript p indicates that distances are computed based on a set of p selected variables $X_1, \cdots, X_p$; in the most common case of the Euclidean distance, we have

$$d_{ij} = \sqrt{\sum_{h=1}^{p} (x_{ih} - x_{jh})^2}$$

with x_{ih} and x_{jh} denoting the value assumed by variable X_h for cases i and j, respectively.

The MDS algorithm aims to display each case in a q-dimensional space ($q << p$) such that the distance matrix $\mathbf{D}^q$ of the obtained configuration fits as closely as possible the dissimilarity matrix Δ or the distance matrix $\mathbf{D}^p$.

Several approaches have been proposed for MDS (Kruskal, 1964a,b; Kruskal and Wish, 1978; Cox and Cox, 2000; Borg et al., 2017). The R function MDSmap optimizes the case locations for a two-dimensional scatter plot ($q = 2$) and is based on Kruskal's nonmetric approach, requiring to minimize the so-called stress index S, a measure of how good is the match between the matrix $\mathbf{D}^q$ of the resulting configuration and the input matrix Δ or $\mathbf{D}^p$. S is a normalized index, so it can be expressed as a percentage. The stress index should be very close to 0. Some authors give rules of thumb for interpreting its value, but we prefer to mention just that it should not exceed 20%. MDS solutions are invariant to rotations and reflections, which consist in transformations of the MDS configuration that leave the distances unchanged. For this reason, the MDS solution is often rotated or reflected in order to make the interpretation easier.

We now show an example of displaying players according to the similarity in their achievements in the following set of 8 variables: scored point, 3- and 2-point shots made, total rebounds (offensive and defensive), assists, turnovers, steals and blocks, limiting attention to those players who have played at least 1500 minutes in the championship

```
> attach(Pbox)
> data <- data.frame(PTS, P3M, P2M, REB=OREB+DREB,
                     AST, TOV, STL, BLK)
> detach(Pbox)
> data <- subset(data, Pbox$MIN>=1500)
> id <- Pbox$Player[Pbox$MIN>=1500]
```

and obtain the two-dimensional configuration displayed in the top panel of Figure 3.4 with the following code lines

```
> mds <- MDSmap(data)
> plot(mds, labels=id)
```

In the map of Figure 3.4, the players are displayed according to their similarity in the selected variables: players close to each other have similar achievements, while a higher distance means more peculiar characteristics. In the top panel map, we notice a big group of players having roughly similar features, some smaller groups of similar players quite separated from the others (*e.g.*, Paul George, Damian Lillard, Kemba Walker, Lou Williams, Kyle Lowry, etc. in the top part of the map and Karl Anthony Towns, Jusuf Nurkic, LaMarcus Aldridge, Rudy Gobert, etc. in the bottom) and a number of players spread in a radial pattern, thus denoting very special features (*e.g.*, James Harden, Russell Westbrook, Lebron James, Ben Simmons, Giannis Antetokounmpo, Anthony Davis, etc.). The goodness of fit is fair ($S = 12.97\%$).

We can highlight some selected players or zoom into specific areas of the map (respectively, middle and bottom panels of Figure 3.4) with the following codes:

```
> selp <- which(id=="Al Horford" | id=="Kyle Korver" |
                id=="Myles Turner" | id=="Kyle Kuzma" |
                id=="Andrew Wiggins")
> plot(mds, labels=id, subset=selp, col.subset="tomato")
> plot(mds, labels=id, subset=selp, col.subset="tomato",
       zoom=c(0,3,0,2))
```

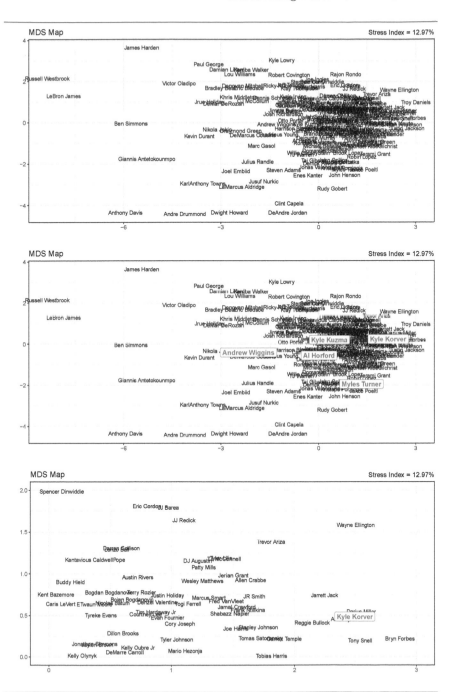

Figure 3.4 Two-dimensional MDS configuration (top: map; middle: map with players highlighted; bottom: map with players highlighted and zoom lens).

Finally, in order to interpret the positioning of players or groups of players in the map with reference to some selected variables (say, $P2M$, $P3M$, AST and REB), we may resort to color-coding the points in the map or, alternatively, overlap to the map a colored "level plot". The former way needs the following code lines and produces the graphs in the top panels of Figure 3.5, where the players are represented as points colored according to the scale reported on the right

```
> plot(mds, z.var=c("P2M","P3M","AST","REB"),
       level.plot=FALSE, palette=topo.colors)
```

The level plot is obtained by fitting the selected variable with a surface, determined with the method of polynomial local regression (Cleveland, 1979; Cleveland and Devlin, 1988) that will be better explained in Chapter 5, Section 5.2.1, with the MDS coordinates as predictors and using second-degree polynomials. It can be obtained by setting `level.plot=TRUE` (that is the default option for this argument). Optionally, we can use `contour=TRUE` in order to draw contour lines. The output is represented in the bottom panels of Figure 3.5.

The joint evaluation of all the graphs helps to determine the meaning of the players' positioning. For example, 2- and 3-point shots made tend to increase when moving from right to left and from bottom to top areas, respectively, so players positioned in the top-left quadrant of the map are quite good on both the achievements (however, the top-performing players in 2- and 3-point shots made tend to be separated, on the left and on the top of the map, respectively). The top-left quadrant is also characterized by a high number of assists. The bottom-left quadrant, instead, is marked by high levels of 2-point shots made and rebounds, but very low levels of 3-points shots made and a rather low number of assists. In the top-right we find an area with high levels of 3-point shots made, but very low levels of 2-points shots made and a rather low performance in assists and rebounds. Finally, the right-bottom quadrant denotes low performance in all four game variables. The analysis can be completed by considering also the other variables, PTS, TOV and STL.

3.4 ANALYZING NETWORK RELATIONSHIPS

Since basketball is a team sport, an important issue is to analyze interactions between players during the game. To this aim, we need to use play-by-play or player tracking data, containing information about events

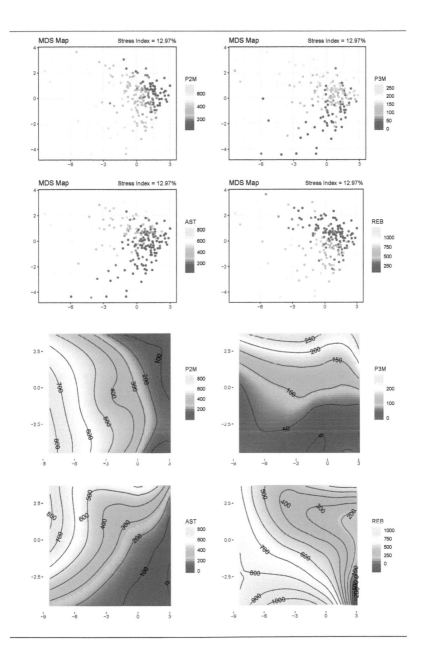

Figure 3.5 Top: Two-dimensional MDS configuration with subjects (points) color-coded according to the variables $P2M$, $P3M$, AST and REB. Bottom: Level plots of the variables $P2M$, $P3M$, AST and REB in the MDS two-dimensional configuration.

occurred during the game. Graphical representations of interactions can be obtained by means of network analysis, which allows to construct and analyze graphs composed of nodes and edges with given attributes, able to represent either symmetric or asymmetric relations between discrete objects. The analysis of passing sequences can be carried out from different perspectives: with reference to the whole path from the in-bounds pass to the shot or just to the last passes (assists and/or the so-called "hockey assists"[3]).

In this section, we describe the R function `assistnet`, designed to investigate the network of assists in a team using play-by-play data. With reference to the Golden State Warriors, we use the play-by-play dataset `PbP`, obtained from `PbP.BDB` thanks to the function `PbPmanipulation` as described on page 23. We start building the network and calculating some assist statistics.

```
> PbP.GSW <- subset(PbP, team=="GSW")
> netdata <- assistnet(PbP.GSW)
```

The object `netdata` of class `assistnet` is a list containing the cross-table of assists made and received by the players (`netdata$assistTable`), a data frame of related statistics (called `netdata$nodeStats`, see Table 3.3) and an object of class `network` (Butts, 2008, 2015) that can be used for further network analyses with specific R packages (`netdata$assistNet`).

A `plot` method is available for this class, so that the command

```
> set.seed(7)
> plot(netdata)
```

returns the graphs in the top panel of Figure 3.6, visualizing the network of assists made and received by the Golden State Warriors players (the arrangement of the nodes can be rotated from one run to another: `set.seed(7)` allows to obtain the same display as ours). In each graph, the players are represented as the network nodes and the oriented edges

---

[3]In ice hockey, there can be two assists per goal, attributed to the last two players who passed the puck towards the scoring teammate. In that context, it has been recognized that the last pass does not account for each and every single way to create a shot opportunity for the teammate, as the whole team's offense has to build a creative passing sequence in order to find the gaps in their opponent's defense. Following this reasoning, also in basketball the idea is taking hold of recording the secondary assist in order to highlight passing skills that otherwise would remain hidden.

TABLE 3.3   Variables of data frame `netdata$nodeStats`.

Variable	Description
FGM	Field goals made
FGM_AST	Field goals made thanks to a teammate's assist
FGM_ASTp	Percentage of FGM_AST over FGM
FGPTS	Points scored with field goals
FGPTS_AST	Points scored thanks to a teammate's assist
FGPTS_ASTp	Percentage of FGPTS_AST over FGPTS
AST	Assists made
ASTPTS	Points scored by assisted teammates

denote the assists made by the player on the startpoint to the teammate on the endpoint. The edge color informs on the number of assists, according to the white-blue-red scale on the right. The default network layout is determined with the Kamada-Kawai algorithm (Kamada and Kawai, 1989), but it can be changed and a threshold can be set in order to display only the edges corresponding to a minimum number of assists (bottom panel of Figure 3.6):

```
> set.seed(7)
> plot(netdata, layout="circle", edge.thr=20)
```

The two nets displayed in Figure 3.6 clearly show that, with respect to the last pass and shot, the hard core of the team is composed by the quartet Stephen Curry, Klay Thompson, Kevin Durant and Draymond Green. Not a surprise, considered that, according to many experts, this foursome has made Golden State Warriors the most recent one of the fabled superteams that NBA happens to offer from time to time. In addition, the peculiarity of these four players with reference to the net assist-shot is the singular way Stephen Curry and Draymond Green interpret their position, with the point guard assisting much less than the power forward.

The network graph shows that Draymond Green is the greatest assist-man of the team, with a clear preference for assisting Thompson and Durant. To a lesser extent, Stephen Curry both makes and receives assists, mainly interacting with the other three. Klay Thompson is by far the most assisted player, not only by the other three of the above-mentioned quartet, but also by several other teammates. Finally, Kevin Durant mainly engages with the other three but exhibits an appreciable number of assists made to Zaza Pachulia, Jordan Bell and JaVale McGee and received by Andre Iguodala.

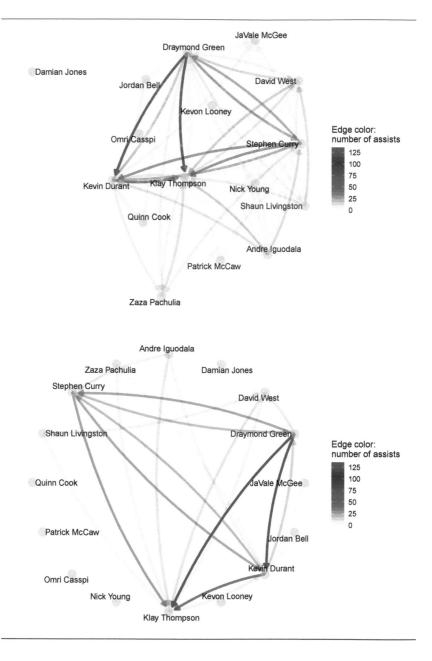

Figure 3.6 Network of assists (layouts without node statistics). Top: basic layout with nodes arranged with Kamada-Kawai layout. Bottom: layout in circle with threshold (= 20) on the edges.

An interesting follow-up on this issue can be obtained by considering the moments when Draymond Green, the most central player with respect to assists made, is not on the field. This allows to investigate how the team reorganizes the balance of its assist-shot passing sequence in absence of the main "bandleader". To do that, we select the rows of PbP that do not contain the name of Draymond Green in the variables a1, ..., a5, h1, ..., h5 (see Table 2.2)

```
> cols <- paste0(c("a","h"), rep(1:5,each=2))
> PbP.GSW.DG0 <- PbP.GSW[!apply(PbP.GSW[,cols], 1, "%in%",
                         x="Draymond Green"),]
> netdata.DG0 <- assistnet(PbP.GSW.DG0)
> set.seed(1)
> plot(netdata.DG0)
```

The resulting network, displayed in Figure 3.7, shows that, absent Draymon Green, the other players tend to share out the work of assisting the teammates. The new central man is Kevin Durant and, to a lesser

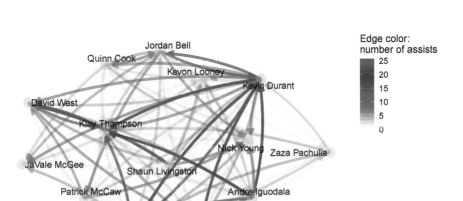

Figure 3.7 Network of assists (layouts without node statistics) of the lineups without Draymond Green.

extent, Andre Iguodala. As mentioned before, although being a point guard, Stephen Curry seems to take a less prominent role from the point of view of the assists.

In this respect, we may be interested in computing the average points and the average play length with and without Draymond Green, in order to assess which is the most efficient game strategy between the two depicted by the networks in Figures 3.6 and 3.7. We first compute average points in 2- and 3-point shots (object p0) and average play length (object pl0) in the game phases without Draymond Green

```
> PbP.GSW.DG0 <- subset(PbP.GSW.DG0,
                    ShotType=="2P" | ShotType=="3P")
> p0 <- mean(PbP.GSW.DG0$points)
> pl0 <- mean(PbP.GSW.DG0$playlength)
```

Similarly, after selecting the game phases with Draymond Green on the court, we obtain the objects p1 and pl1

```
> PbP.GSW.DG1 <- PbP.GSW[apply(PbP.GSW[,cols], 1, "%in%",
                    x="Draymond Green"),]
> PbP.GSW.DG1 <- subset(PbP.GSW.DG1,
                    ShotType=="2P" | ShotType=="3P")

> p1 <- mean(PbP.GSW.DG1$points)
> pl1 <- mean(PbP.GSW.DG1$playlength)
```

The results are

```
> p0
[1] 1.108733
> pl0
[1] 11.22426
> p1
[1] 1.159812
> pl1
[1] 10.82656
```

When Draymond Green is on the court, we have on average a shorter play length (10.8 versus 11.2 seconds) and more efficient field goals (1.16 versus 1.11 points per shot), which means a faster pace with more points scored.

The **plot** method allows to specify some statistics (chosen among those available in the data frame **netdata$nodeStats**) for the node size and color, thus adding relevant information to the graph. For example, going back to the network built on the complete dataset, the following code lines allow us to obtain the nets displayed on top and bottom of Figure 3.8, respectively.

```
> plot(netdata, layout="circle", edge.thr=20,
        node.col="FGPTS_AST", node.size="ASTPTS")
> plot(netdata, layout="circle", edge.thr=20,
        node.col="FGPTS", node.size="FGPTS_ASTp")
```

The top panel of Figure 3.8 shows a net with nodes characterized by the number of points scored thanks to teammates' assists (node color) and the number of points scored by teammates thanks to the player's assists (node size). For example, Draymond Green has a large blue circle, denoting that he creates scoring opportunities for teammates more than scoring points himself thanks to teammates' assists. Quite the opposite, Klay Thompson exhibits a small red circle, due to a high number of assisted scored points and a low number of assists made. In the bottom panel of Figure 3.8, the nodes are characterized by the total number of points scored with field goals (node color) and the percentage of these points scored thanks to received assists (node size). Here it is interesting to note the small red circle of Stephen Curry, denoting a player who scores much but he often creates his own shot (53.7% assisted shots). Quite the same can be said for Kevin Durant (56.0%). On the other hand, Klay Thompson's scored points are for the most assisted by teammates (84.7%). The other players tend to have mostly large circles, meaning that the points they score are often assisted (for example, JaVale McGee 82.4%, Zaza Pachulia 81.9%, Omri Casspi 81.6%, Nick Young 81.3%).

The information contained in the object netdata can be used to perform further analyses, exploiting the basic functions presented in Chapter 2. We first extract the cross-table of made/received assists and the data frame of additional statistics,

```
> TAB <- netdata$assistTable
> X <- netdata$nodeStats
```

then we merge X with the Pbox data frame in order to incorporate potentially useful available information

```
> names(X)[1] <- "Player"
> data <- merge(X, Pbox, by="Player")
```

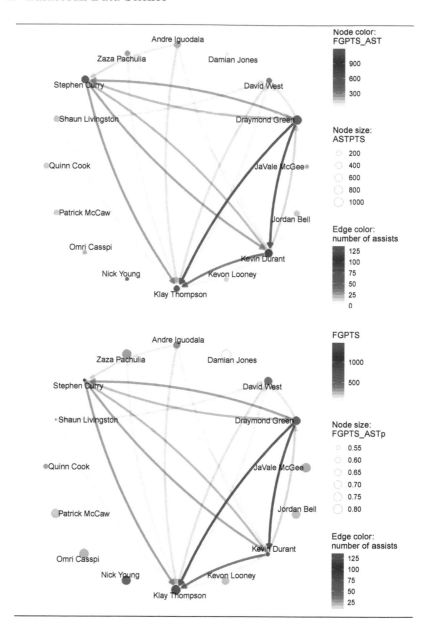

Figure 3.8 Network of assists (layouts with node statistics). Top: nodes with FGPTS_AST (Points scored thanks to a teammate's assist; node color) and ASTPTS (Points scored by assisted teammates; node size). Bottom: nodes with FGPTS (Points scored with field goals; node color) and FGPTS_ASTp (Percentage of points scored thanks to a teammate's assist over points scored with field goals; node size).

We can resort to the function `scatterplot` described in Chapter 2, Section 2.2.4 in order to investigate more deeply the issue of the percentage of assisted shots. We generate the scatter plot of players according to the number of field goals made and the percentage of assisted ones, with color coding determined by the minutes played (Figure 3.9). It clearly shows how assisted are the players, how much they score and how long they play.

```
> mypal <- colorRampPalette(c("blue","yellow","red"))
> scatterplot(data, data.var=c("FGM","FGM_ASTp"),
              z.var="MIN", labels=data$Player,
              palette=mypal, repel_labels=TRUE)
```

Another interesting issue is to investigate, for both assists made and received, whether a player tends to assist (receive assists from) a small or a large number of teammates. In other words, we can analyze the inequality (see Chapter 2, Section 2.2.7) of assists made and received.

We first select the players who have played at least one quarter per game on average (> 984 minutes)

```
> sel <- which(data$MIN>984)
> tab <- TAB[sel,sel]
```

Subsequently, for each selected player, we exploit the function `inequality` to draw the Lorenz curve and compute the Gini coefficient

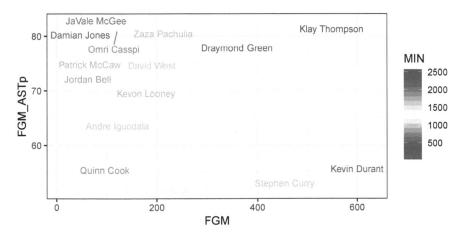

Figure 3.9 Scatter plot of field goals made (x-axis) vs. percentage of assisted field goals (y-axis); color = minutes played.

of assists made and received (respectively, the rows and columns of `tab`). To do that, we use the following code lines with a `for` loop

```
> no.pl <- nrow(tab)
> pR <- pM <- vector(no.pl, mode="list")
> GiniM <- array(NA, no.pl)
> GiniR <- array(NA, no.pl)
> for (pl in 1:no.pl) {
        ineqplM <- inequality(tab[pl,], npl=no.pl)
        GiniM[pl] <- ineqplM$Gini
        ineqplR <- inequality(tab[,pl], npl=no.pl)
        GiniR[pl] <- ineqplR$Gini
        title <- rownames(tab)[pl]
        pM[[pl]] <- plot(ineqplM, title=title)
        pR[[pl]] <- plot(ineqplR, title=title)
        }
```

In the end, we arrange all the plots in a unique frame

```
> library(gridExtra)
> grid.arrange(grobs=pM, nrow=2)
> grid.arrange(grobs=pR, nrow=2)
```

The resulting graphs are displayed in Figure 3.10, showing appreciable differences among players from the point of view of their interactions with teammates in the last pass and shot. We get an easier understanding of results by inspecting the values of `tab` (Table 3.4) in tandem with the graphs.

For what concerns the assists made (top panel of Figure 3.10), Stephen Curry is the one with the highest inequality level (Gini index =

TABLE 3.4 Assists made (rows) and received (columns) by the selected players.

	AI	DW	DG	KD	KT	NY	SL	SC
Andre Iguodala (AI)	0	18	20	30	37	20	9	34
David West (DW)	8	0	21	12	38	15	19	2
Draymond Green (DG)	18	36	0	125	132	21	24	66
Kevin Durant (KD)	7	14	44	0	83	15	13	56
Klay Thompson (KT)	14	31	18	30	0	4	11	17
Nick Young (NY)	0	5	1	3	7	0	1	3
Shaun Livingston (SL)	6	33	19	10	27	18	0	6
Stephen Curry (SC)	14	3	54	65	68	20	1	0

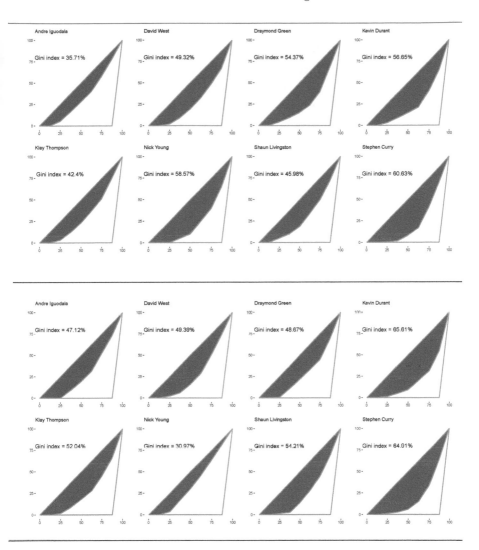

Figure 3.10 Lorenz curves (with Gini indexes) of players for assists made (top) and received (bottom).

60.63%), due to the fact that he mostly assists Klay Thompson (68 assists), Kevin Durant (65) and Draymond Green (54). Other players with high inequality in assists made are Nick Young (Gini index = 58.57%, but he only made 20 assists), Kevin Durant (56.65%) and Draymond Green (54.37%). Diametrically opposed to them, Andre Iguodala exhibits the highest balance (Gini index = 35.71%), as he tends to equally assist all the teammates.

On the side of the assists received (bottom panel of Figure 3.10) the highest inequality levels are again up to Kevin Durant (Gini index = 65.61%) and Stephen Curry (64.91%), due to the fact that both of them are mostly assisted by Draymond Green (125 and 66 assists from Green to Durant and Curry, respectively) and they also assist each other (65 and 56 assists, respectively). The highest balanced situation is that of Nick Young (Gini index = 30.97%) who is almost equally assisted by all the teammates.

The association measures introduced in Section 3.1.1 between assists made and received, as reported in Table 3.4, can be calculated with the following code lines

```
> library(vcd)
> assocstats(tab)
```

Results show a highly significant association between the variables assists made and received ($X^2 = 507.67$, with p-value= 0; Cramer's V equals 0.23 and Pearson's contingency coefficient is $P = 0.512$). This suggests that there is a statistical dependence between assists made and received and, therefore, there are pairs of players who tend to interact much more (or less) than what we should expect if the choice of the teammate to assist was random.

A graphical summary of assists made and received by the selected players, from the perspective of both number and inequality, can be obtained thanks to a bubble plot (Chapter 2, Section 2.2.5). We first compose the data frame XX with all the necessary variables, then we resort to the function bubbleplot

```
> XX <- data.frame(X[sel,], GiniM, GiniR)
> labs <- c("Gini Index for assists made",
            "Gini Index for assists received",
            "Assists received", "Assists made")
> bubbleplot(XX, id="Player", x="GiniM", y="GiniR",
             col="FGM_AST", size="AST",
             labels=labs, text.size=4)
```

Remember that, as mentioned in Section 2.2.5, the bubble size is rescaled between 0 and 100 and the rescaling can be disabled setting the argument scale.size to FALSE. The graph of Figure 3.11 shows that Kevin Durant and Stephen Curry interact with a limited number of teammates both

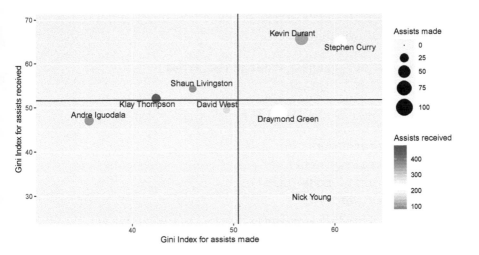

Figure 3.11 Bubble plot with Gini index of assists made (x-axis) and received (y-axis), number of assists made (size), number of assists received (color).

for assists made and received. They make a quite high number of assists, but Kevin Durant receives much more than Stephen Curry. Draymond Green is the one with the biggest bubble and he makes more assists, quite concentrated to a limited number of teammates, than he receives. David West and Shaun Livingston occupy a middle position for all the variables. More noticeable is the situation of Klay Thompson, who receives much more assists than he makes and tends to engage with a high number of teammates both for assists made and received. The most balanced player with respect to assists-shot interactions is Andre Iguodala, who, however, has a quite low number of assists made and received.

Further investigation with network analysis tools such as central-ity measures (*e.g.*, centrality degree, betweenness centrality, closeness centrality, etc.) is also possible, thanks to proper R packages, which require specific objects in input. These objects can easily be derived starting from the object of class `assistnet`. For example, we may refer to the libraries `tidygraph` (Pedersen, 2019), `igraph` (Csardi and Nepusz, 2006), and `CINNA` (Ashtiani, 2019):

```
> library(tidygraph)
> library(igraph)
> library(CINNA)
```

First, the data frame `netdata$assistNet` is transformed into an object suitable for network analysis

```
> net1 <- as_tbl_graph(netdata$assistNet)
> class(net1) <- "igraph"
```

then, a lot of metrics are available for easy computation

```
> centr_degree(net1)
> alpha_centrality(net1)
> closeness(net1, mode="all")
> betweenness(net1)
> calculate_centralities(net1)
```

The correct use and understanding of these metrics require expertise in network analysis beyond the scope of this manual. To those interested in deepening this issue, we recommend a good text on this topic (for example, Newman, 2018).

3.5 ESTIMATING EVENT DENSITIES

3.5.1 Density with respect to a concurrent variable

An important issue for the analysis of games is to determine the frequency of occurrence of some events with respect to some other concurrent variable. To this aim, we can resort to density estimation methods, a wide set of tools such as histograms, naive, kernel, nearest neighbour, variable kernel, orthogonal series, maximized penalized likelihood and general weight function estimators (see Silverman, 2018 for a comprehensive discussion).

In the Foreword, Ettore Messina claimed that, "... *scoring ten points in the first two quarters of the game is not the same as scoring them in the final minutes...*". So, we may be interested in assessing the pattern of the shooting frequency in time or space. The R function `densityplot` requires play-by-play data to compute and plot kernel density estimates of shots with respect to the period time (`periodTime`, seconds played in a given quarter), the total time (`totalTime`, seconds played in total), the play length (`playlength`, time between the shot and the immediately preceding event), the shot distance (`shot_distance`, distance in feet from the basket). These variables have been chosen keeping in mind the importance of investigating, other than average performance, players' behavior in specific situations, as suggested by coach Messina.

densityplot uses a Gaussian kernel density estimation procedure and a bandwidth computed by default as the standard deviation of the kernel, but adjustable by the researcher. A kernel is a window function used in nonparametric statistics to define a weighting system for data lying around a point at which a function is estimated. The bandwidth allows to control the window width. Several types of kernel functions are available, the most commonly used are uniform, triangle, Gaussian, quadratic and cosine. We will deal again with the concept of kernel in Chapter 5, Section 5.2, in the context of nonparametric regression.

To give an example of density estimation, we may be interested in the density of 2-point shots of the Golden State Warriors (Figure 3.12) and all their opponents (Figure 3.13) with respect to all four available variables (period time, total time, play length, shot distance). First, we select, from the play-by-play dataset PbP (obtained from PbP.BDB thanks to the function PbPmanipulation as described on page 23), the rows describing a shot (PbP$result != "") attempted by the team (data.team) or its opponents (data.opp).

```
> data.team  <- subset(PbP, team=="GSW" & result!="")
> data.opp <- subset(PbP, team!="GSW" & result!="")
```

Then, we obtain the graphs with the following code lines, where the argument shot.type allows to decide which type of shots we wish to analyze (here 2P, but 3P, FT, field are also available[4]).

```
> densityplot(data=data.team, shot.type="2P",
              var="periodTime", best.scorer=TRUE)
> densityplot(data=data.team, shot.type="2P",
              var="totalTime", best.scorer=TRUE)
> densityplot(data=data.team, shot.type="2P",
              var="playlength", best.scorer=TRUE)
> densityplot(data=data.team, shot.type="2P",
              var="shot_distance", best.scorer=TRUE)
> densityplot(data=data.opp, shot.type="2P",
              var="periodTime", best.scorer=TRUE)
> densityplot(data=data.opp, shot.type="2P",
              var="totalTime",best.scorer=TRUE)
> densityplot(data=data.opp, shot.type="2P",
              var="playlength", best.scorer=TRUE)
```

---

[4]The option FT is available only with the variables' total time and period time.

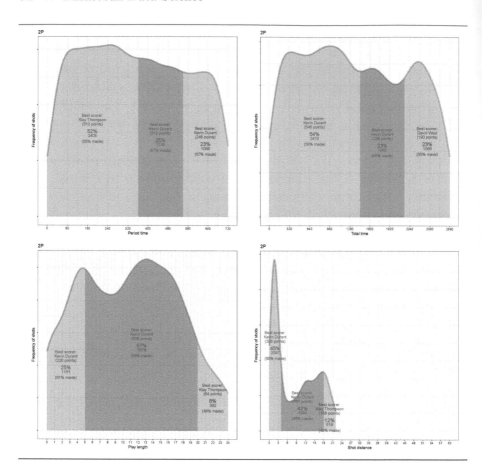

Figure 3.12 Density estimation of 2-point shots of the Golden State Warriors, with respect to period time, total time, play length, and shot distance.

```
> densityplot(data=data.opp, shot.type="2P",
              var="shot_distance", best.scorer=TRUE)
```

The curve in the plots shows the estimated densities and, with text, details about three selected areas (by default: number/percentage of shots and percentage of shots made in the corresponding interval; with the additional option best.scorer=TRUE: player who scored the highest number of points in the corresponding interval). The range boundaries of the three regions have been set to the following values: for period time and total time in Figures 3.12 and 3.13, they refer to half and 3/4 of time; boundaries for play length are chosen according to results

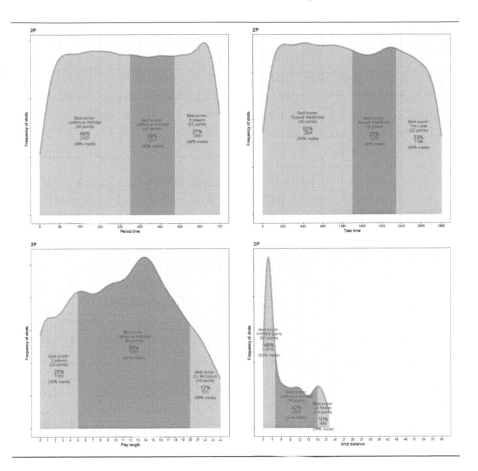

Figure 3.13 Density estimation of 2-point shots of the Golden State Warriors' opponents, with respect to period time, total time, play length, and shot distance.

of the FOCUS study (see end of chapter), while for the shot distance they indicate the restricted area arc (4 feet) and an imaginary line very close to the 3-point line (18 feet)[5]. These default settings that can be adjusted by the researcher with the parameter threshold (see the help of densityplot).

Comparing Figure 3.12 and Figure 3.13, several insights can be gained about the 2-point shot behavior of the Golden State Warriors'

---

[5]These boundaries are relative to 2-point shots; in the case of 3-point shots, the boundaries are 25 and 30 feet, while for field shots they are set at 4 and 22 feet.

players and their opponents. For example, the players of the Golden State Warriors:

- tend to concentrate their shots in the first half of each quarter (52% against 49% of the opponents) and in general in the first half (54% against 52% of the opponents) of the whole game (however, these percentages do not differ much from 50%, so these evidences are rather weak)

- tend to shoot very fast (25% of shots in the first 5 seconds after the previous event, against 23% of the opponents) and quite rarely are forced to shoot close to the 24-second buzzer sound (8% against 12% of the opponents), but also in this situation they retain a very high scoring performance (49% against 39% of the opponents)

- "take it to the rim" (shot distance less than 4 feet, *i.e.*, from the restricted area arc) more rarely than their opponents (45% against 48%), but they retain very high scoring performance when shooting from higher distance (46% even from very close to the 3-point line, against 36-37% of the opponents).

Several other remarks can be drawn by considering the best scorers in the different situations and other kinds of shots, setting the argument shot.type.

In addition, it can be interesting to inspect single player performances. For example, focusing attention on the shot behavior of Kevin Durant and Stephen Curry with respect to play length and shot distance, we write

```
> KD <- subset(PbP, player=="Kevin Durant" & result!="")
> SC <- subset(PbP, player=="Stephen Curry" & result!="")
> densityplot(data=KD, shot.type="field",
              var="playlength")
> densityplot(data=KD, shot.type="field",
              var="shot_distance")
> densityplot(data=SC, shot.type="field",
              var="playlength")
> densityplot(data=SC, shot.type="field",
              var="shot_distance")
```

and obtain the graph of Figure 3.14, showing the very different way of playing of these two outstanding players. Specifically, we see that Kevin

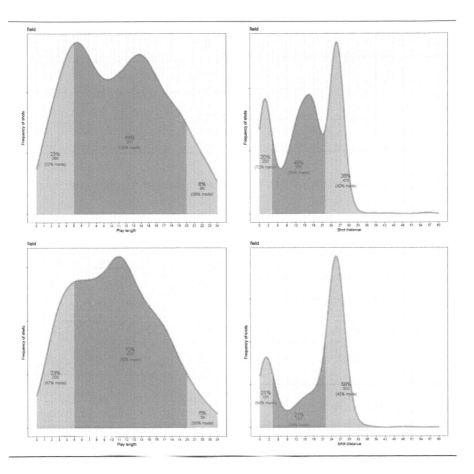

Figure 3.14   Density estimation of field shots of Kevin Durant (top panels) and Stephen Curry (bottom panels), with respect to play length and shot distance.

Durant tends to shoot more often close to the 24-second buzzer sound (8% against 5% of Stephen Curry), although he has a significantly lower performance in this situation (39% against 53%). However, he tends to perform better in the first 20 seconds of the play (52-53% against 47-50%). For what concerns the shot distance, they both attempt 20% of shots from the restricted area, with Durant being far more effective (72% made, versus only 64% for Curry). From the middle distance, Curry performs better (55% against 50% of Durant), but shoots less (only 21% of shots against 45%). The majority (58%) of Curry's shots is from the long distance, with about the same performance of Durant.

3.5.2 Density in space

When the frequency of occurrence of the events is estimated with respect to variables denoting space coordinates, we resort to spatial analysis tools and graphical solutions, such as density polygons, raster, and hexbins. For the estimation of shot density across the court, we can refer to the function shotchart, introduced in Chapter 2, Section 2.2.8, that allows to specify different types of shot density charts by opportunely setting the option type. We need to use play-by-play data with adjusted shot coordinates, with the same linear transformation used in Chapter 2, Section 2.2.8:

```
> PbP$xx <- PbP$original_x/10
> PbP$yy <- PbP$original_y/10 - 41.75
```

Let us be interested in analyzing the spatial density of Klay Thompson's shots. We first select his shots

```
> KT <- subset(PbP, player=="Klay Thompson")
```

Then, we can plot spatial density estimates using different options for the argument type:

```
> shotchart(data=KT, x="xx", y="yy",
            type="density-polygons")
> shotchart(data=KT, x="xx", y="yy", type="density-raster")
> shotchart(data=KT, x="xx", y="yy", type="density-hexbin")
```

Additional options are also available, in order to customize the color palette, decide the number of hexbins, add the scatter of shots:

```
> shotchart(data=KT, x="xx", y="yy",
            type="density-polygons", scatter=TRUE)
> shotchart(data=KT, x="xx", y="yy", type="density-raster",
            scatter=TRUE, pt.col="tomato", pt.alpha=0.1)
> shotchart(data=KT, x="xx", y="yy", type="density-hexbin",
            nbins=50, palette="bwr")
```

The six graphs obtained with the above code lines are displayed in Figure 3.15, where we clearly see that Klay Thompson tends to shoot from both the restricted area and the middle-long distance. In the latter case, he seems to prefer moving to his left-hand side.

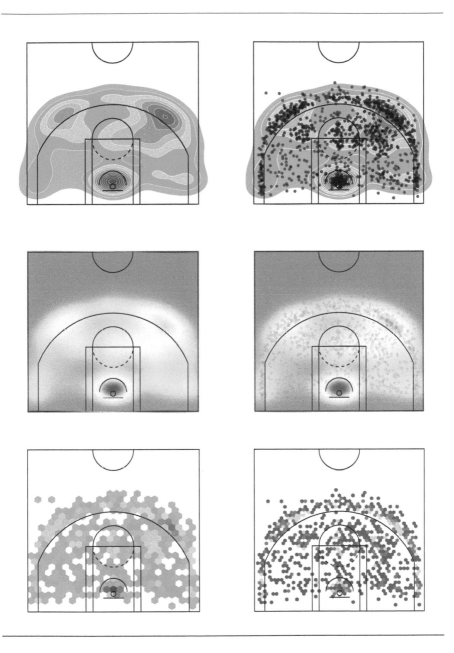

Figure 3.15 Spatial density estimation of Klay Thompson's shots with basic (left panels) and additional (right panels) options.

3.5.3 Joint density of two variables

A last case of density estimation is related to the correlation analysis between variables, using the function `scatterplot` introduced in Chapter 2, Section 2.2.4, and again used in Section 3.2 to analyze pairwise linear correlation of a set of variables. Let us select those players who have played at least 500 minutes and consider the following variables: scored point, 3- and 2-point shots made, total rebounds (offensive and defensive) and assists (per minute played):

```
> data <- subset(Pbox, MIN>=500)
> attach(data)
> X <- data.frame(PTS, P3M, P2M, REB=OREB+DREB, AST)/MIN
> detach(data)
```

We can plot a correlation matrix with univariate density estimates on the diagonal and contour estimates of the pairwise joint densities in the lower triangle.

```
> scatterplot(X, data.var=1:5,
            lower=list(continuous="density"),
            diag=list(continuous="densityDiag"))
```

The outcome is shown in Figure 3.16.

3.6 FOCUS: SHOOTING UNDER HIGH-PRESSURE CONDITIONS

A broad set of extremely powerful tools to discover hidden structures and patterns in data comes from machine learning techniques that make a large number of algorithms available for both supervised and unsupervised learning. The philosophy behind algorithmic modeling is discussed by Breiman (2001b) and a number of machine learning algorithms are well addressed in Friedman et al. (2009) and Witten et al. (2016). In this section, we show some results obtained by means of CART (Classification And Regression Trees, Breiman et al., 1984), an algorithm able to discover the way a set of variables affect a given outcome, taking into account possibly nonlinear or complex relationships and interactions among the predictors, discussed in a paper aimed at describing the influence of some high-pressure game situations on the shooting performance (Zuccolotto et al., 2018).

In the Foreword, Ettore Messina claimed that *"...the coach's decision as to who to put on the field is based above all on feelings about*

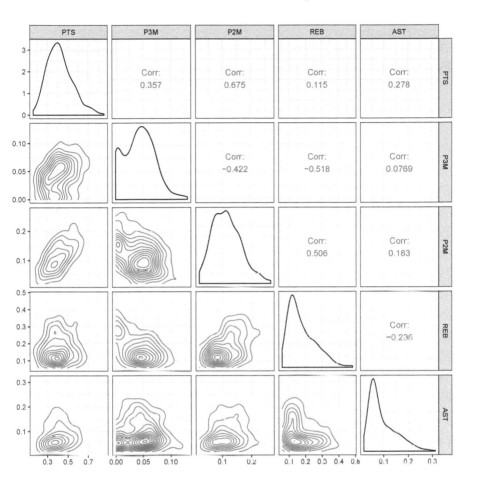

Figure 3.16 Correlation matrix with univariate density estimates on the diagonal and joint density contours in the lower triangle.

the character of the individual players and their personal chemistry with the teammates, rather than exclusively their technical skills, about their ability to cope with pressure." Following the necessity to try to measure players' ability to cope with pressure, this FOCUS study concentrates on shooting performance and moves from the idea that the shots are not all the same, as they can occur in (relatively) quiet or high-stress situations. Preliminary discussions about the definition of high-pressure game situation and its essential meaning, both from a psychological and competition-related perspective, can be found on the cited paper, which also contains some additional issues concerned with (i) the analysis of univariate relations between single variables and scoring probabilities and (*ii*) the assessment of players' personal reactions to some selected high-pressure game situations. Issue (*i*) makes use of nonparametric regression via Gaussian kernel smoothing and will be discussed in Chapter 5, Section 5.2.2.

In this section we only deal with the main aim of the paper, consisting in

- identifying the game situations that, generating pressure on the players (and beyond the obvious fact that different players can have different reactions to this pressure), turn out to be associated to lower scoring probabilities,

- showing how these findings can be used to develop new shooting performance measures, improving upon currently employed shooting statistics.

The analysis has been performed on play-by-play data of all matches played during the Italian Serie A2 championship 2015/2016 (the second tier of the Italian league pyramid, just below the first division, Serie A). Data contain almost 70,000 shots, so the sample size is large enough to guarantee robust estimates of the scoring probabilities, even in situations that may occur only occasionally. To take into account that the reaction to pressure may be different according to the professional level of the players (Madden et al., 1990, 1995), the most important results have then been checked on a smaller play-by-play dataset from the Olympic Basketball Tournament Rio 2016. Table 3.5 summarizes the main features of the two datasets.

TABLE 3.5 Main features of the datasets for the two case studies.

Dataset	"Serie A2"	"Rio 2016"
Competition	Championship - regular season	Olympic Tournament
Period	2015, 4th Oct - 2016, 23rd Apr	2016, 6th - 21st Aug
Gender	Male	Male
Matches	480	38
Teams	32	12
Players	438	144
2-point shots	33682 (50.9% Made)	3101 (52.2% Made)
3-point shots	21163 (34.1% Made)	1780 (33.8% Made)
Free throws	14843 (73.5% Made)	1589 (74.8% Made)

Source: Zuccolotto et al. (2018)

The authors identified a game situation as being "high-pressure" if it is for some reason more troublesome and demanding, without trying to distinguish whether the pressure comes from game-related factors, psychological factors, or both. Following this general definition and considering the available data, they identified, also following some suggestions of basketball experts, some main types of situations that may generate pressure on the player when a shot is attempted: when the shot clock is going to expire, when the score difference with respect to the opponent is small, when the team as a whole has performed poorly during the match up to that particular moment in the game, when the player who is shooting has missed his previous shot, and when the time left on the clock is running out. The authors used CART to assess the impact of these high-pressure situations on the scoring probability by taking account of all the joint associations among the variables[6]. The authors preferred to use categorical covariates instead of numerical ones, as this prevents trees from growing too deeply (thus protecting from instability, the tendency of trees to appreciably change their structure when little differences are impressed on the learning step) and allows focusing only on the most interesting situations. To convert the numerical covariates into categorical ones, the authors suggested identifying the cut-off values by combining the results of a machine learning procedure and some basketball experts' suggestions. As for the machine learning procedure, they proposed a procedure to obtain a *threshold importance measure*, computed by growing a preliminary tree using the numerical covariates and then summing up all the decreases of the heterogeneity index allowed by each threshold of each covariate in all of the tree nodes. The results are displayed in Figure 3.17 for the variables denoting the time seconds until the 24-second buzzer sound (Shot.Clock), the time to the end of each quarter (Time), the fraction of missed shots for the whole team up to the moment when each shot is attempted (Miss.T), the score difference with respect to the opponent when each shot is attempted (Sc.Diff).

Some remarks can be drawn based on the graphs of Figure 3.17:

- the two datasets, although very different from the point of view of the players' professional level, have given very similar results;

- the obtained results are consistent with preliminary experts' suggestions about what the threshold values should be;

---

[6]The analysis with CART can be easily carried out by using the R packages `tree` and `rpart`. Nice plotting options are allowed by the package `rpart.plot`.

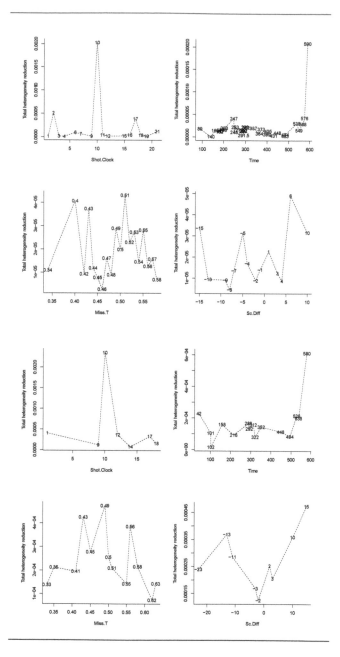

Figure 3.17 Threshold importance measures, dataset "Serie A2" (top) and "Rio 2016" (bottom). Source: Zuccolotto et al. (2018).

- variable Shot.Clock presents some clear thresholds at the extremes and in the middle (which is consistent with the experts' suggestions of isolating the extremes);

- for variable Time, the thresholds' importance begins to grow in the last 100 seconds (which is also consistent with the experts' opinions regarding the relevance of the last 1-2 minutes of each quarter);

- variable Miss.T presents several thresholds between 0.35 and 0.60, but we do not recognize peaks that may suggest a possible best choice within this interval, so the split can be made according to percentiles;

- variable Sc.Diff also presents several different peaks. Among them, we recognize the possibility of following the experts' suggestions to isolate low score differences (approximately between -5 and 1), but we also find some possible thresholds at values ≤ -10 and ≥ 6.

On the whole, the authors decided to convert numerical variables into categorical variables according to the criteria summarized in Table 3.6.

It is worth pointing out that the variable Shot.Clock describes the seconds until the buzzer sounds (*i.e.*, the value displayed on the shot clock, regardless of whether the shot clock had been previously reset

TABLE 3.6 Conversion into categorical covariates.

Shot.Clock	early: Shot.Clock > 17 early-middle: $10 <$ Shot.Clock ≤ 17 middle-end: $2 <$ Shot.Clock ≤ 10 time-end: Shot.Clock ≤ 2
Time	normal: Time ≤ 500 quarter-end: Time > 500
Miss.T	Bad: Miss.T $\leq$ 25th percentile Medium: 25th percentile $<$ Miss.T $\leq$ 75th percentile Good: Miss.T $>$ 75th percentile
Sc.Diff	less than -15: Sc.Diff ≤ -15 between -15 and -5: $-15 <$ Sc.Diff ≤ -5 between -5 and 1: $-5 <$ Sc.Diff ≤ 1 between 1 and 6: $1 <$ Sc.Diff ≤ 6 more than 6: Sc.Diff > 6

Source: Zuccolotto et al. (2018)

to 14 seconds). As a matter of fact, the extra time with respect to the 24 seconds may temporarily modify the pressure of the game and affect the scoring probability. For this reason, the authors introduced in the CART an additional categorical predictor (Poss.type) denoting whether the shot is made during the original 24 seconds on the shot clock or after the shot clock has been reset to 14 seconds.

The CART model with all the categorical covariates was grown using the Gini index as the split selection criterion. After pruning, the trees shown in Figure 3.18 have been obtained.

The trees provide very interesting interpretations about the impact of high-pressure game situations on the probability of scoring. The first splits are made according to the variable Shot.Type (2-point and 3-point shots to the left branch and free throws to the right). After that, Shot.Clock has the most prominent role for field shots, immediately followed by Sc.Diff, Time and Poss.type, which seem to play a role in interactions with Shot.Clock and among them. For free throws, the relevant variable is Miss.Pl (denoting whether the previous shot of the shooting player scored a basket or not).

The tree based on the "Serie A2" dataset is the most deeply grown, thanks to its higher sample size. In detail, this tree reveals the following relationships:

- for free throws, the only relevant variable is Miss.Pl: the estimated scoring probability when the previous shot by that player scored a basket is 0.7481 versus 0.7115 when the previous shot was missed;

- for 3-point shots, the first relevant variable is Shot.Clock: the estimated scoring probability at time-end is 0.2939; otherwise, the scoring probability is 0.3504 and 0.3844 for middle-end and earlier shots, respectively, provided the game is not in quarter-end, when the scoring probability decreases to 0.3119;

- for 2-point shots, the most relevant variable is again Shot.Clock: the estimated scoring probability is 0.4086, 0.4764, and 0.6580 for time-end, middle-end, and early shots, respectively; for early-middle shots, we distinguish between the game played after the shot clock has been reset to 14 seconds (*i.e.*, in the first 4 seconds after the shot clock resetting, when the scoring probability increases to 0.6035) and 24-second possessions (a scoring probability of 0.4678 when the score difference is less than -15 and a probability of 0.5501 otherwise).

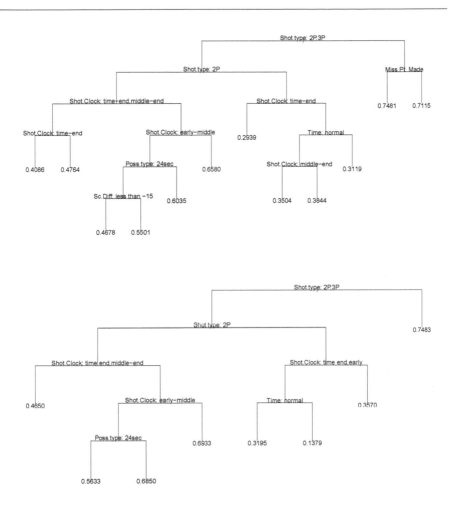

Figure 3.18 Final CART models, dataset "Serie A2" (top) and "Rio 2016" (bottom). Source: Zuccolotto et al. (2018). Each split is made according to variable: threshold indicated (move to the left branch when the stated condition is true); values in the leaves (terminal nodes) are the predicted class values, that is the estimated scoring percentages of the shots belonging to that class.

Very similar remarks, although less detailed since the growth of the tree is less deep, can be drawn from the second CART, obtained with the "Rio 2016" dataset.

According to the evidence identified by the CART models, shots are not all alike. For this reason, the authors used the trees to develop new shooting performance measures, taking into account the circumstances under which each shot has been attempted. For example, a 2-point shot attempted in the last 2 seconds of the shot clock has a scoring probability of approximately 40%, unlike a shot attempted in the first 7 seconds, which has a scoring probability greater than 65%: a shooting performance measure should take into account this evidence and give higher merit to a basket made when the scoring probability is lower. For each shot type T (2P: 2-point, 3P: 3-point, FT: free-throw), let J_T be the set of attempted shots of type T and x_{ij} the indicator assuming value 1 if the jth shot of the ith player scored a basket and 0 otherwise. The new shooting performance of player i for shot type T is given by

$$P_i(T) = av_{j \in J_T} \left(x_{ij} - \pi_{ij} \right) \tag{3.1}$$

where $av_{j \in J_T}(\cdot)$ denotes averaging over all of the shots of type T attempted by player i and π_{ij} is the scoring probability assigned by the CART model to the jth shot of the ith player, that is to a shot of the same type and attempted in the same game situation as the jth shot of the ith player. For each shot, the difference $x_{ij} - \pi_{ij}$ can be used as a performance measure of the shot. In fact, the difference is positive if the shot scored a basket (and the lower the scoring probability, the higher its value) and negative if it missed (and the higher the scoring probability, the higher its absolute value). For example, a basket is worth more when the scoring probability of the corresponding shot is low, whereas when a miss occurs, it is considered more detrimental when the scoring probability of the corresponding shot is high. The value of P_i can be interpreted depending on whether it is positive (meaning a positive balance between made and missed shots, taking into account their scoring probabilities) or negative, and considering its absolute value. The highest interpretability is reached when players are compared by means of a bubble plot with $P_i(2P)$ and $P_i(3P)$ on the x-axis and y-axis, respectively, and the color representing $P_i(FT)$. Figure 3.19 refers to the dataset "Rio 2016" and displays only players who attempted at least 15 shots for each shot type.

In the bubble plot, we can see which players are better (or worse) than the average, for each shot type and taking into account the particular game situation. For example, Rodolfo Fernandez performs better than the average both on 2-point shots and 3-point shots but worse

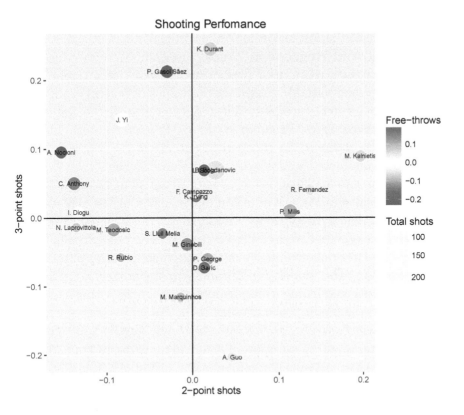

Figure 3.19 Bubble plot of the players' shooting performance measures ("Rio 2016" dataset). Source: Zuccolotto et al. (2018).

on free throws. Most importantly, the graph allows us to see some difference among players who have the same performance, based on the usual goal percentages, where each shot is considered to be the same as any other. For example, Bojan Bogdanovic and Andrés Nocioni have approximately the same 3-point field goal percentages (45% and 45.2%); however, if we consider the difficulty of the game situations in which they attempted their shots, Nocioni performs 3-point shots slightly better than Bogdanovic. In addition, we can also find some players with different field goal percentages that receive equal evaluations when the difficulty of the game situation enters the performance measure. For example, Nicolás Laprovittola, Carmelo Anthony and Ikechukwu Diogu receive a very similar performance measure in 2-point shots, although they have different goal percentages (36.8%, 38.5% and 39.1%, respectively).

GLOSSARY

Classification And Regression Trees: Decision Trees are used in data mining with the aim of predicting the value of a target (outcome, dependent variable) based on the values of several input (predictors, independent variables). The CART procedure (Classification And Regression Trees, Breiman et al., 1984) refers to both Classification Trees (designed for categorical outcomes, *i.e.*, taking a finite number of unordered values, with prediction error usually measured in terms of misclassification cost) and Regression Trees (designed for numerical outcomes, *i.e.*, taking continuous or discrete values, with prediction error typically measured by the squared difference between the observed and predicted values). The procedure is based on the concept of heterogeneity of the outcome, broadly referring to its variability. The models are obtained by recursively partitioning the data space and fitting a simple prediction model within each partition (*e.g.*, the mode or the average if the outcome is categorical or numerical, respectively), so the objects are repeatedly split into two parts according to the criterion of achieving maximum homogeneity within the new parts or, correspondingly, to impress the maximum heterogeneity reduction of the outcome through the split. Heterogeneity reduction is measured by means of impurity measures such as the Gini impurity or the variance reduction, according to whether the outcome is categorical or numerical, respectively. As a result, this binary partitioning procedure can be represented graphically as a decision tree. Some data mining techniques, called ensemble methods (Breiman, 1996; Friedman and Popescu, 2008), improve the power of trees by constructing more than one decision tree according to some randomization rule and then obtain the final prediction by a linear combination of the predictions of all the trees. Popular examples are the Random Forest technique (Breiman, 2001a) or the tree-based Gradient Boosting Machine (Friedman, 2001).

Density estimation: In probability and statistics, the term density estimation refers to a set of techniques aimed to obtain an estimate of the probability density function according to which a population is distributed; the observed data used to compute estimates are assumed to be a random sample from that population.

Dissimilarity measure: A dissimilarity δ is a measure computed on each pair (x, y) of elements of a set S, indicating how different the two elements are from each other. Dissimilarities are distance-like measures (see "Distance metrics" below), in the sense that they are usually obtained by means of a mathematical function that has to satisfy relaxed conditions with respect to distance metrics. In the most common definitions, the conditions to be satisfied are only non-negativity ($\delta(x, y) \geq 0$), identity ($\delta(x, y) = 0 \Leftrightarrow x = y$) and symmetry ($\delta(x, y) = \delta(y, x)$), but some authors point out than even symmetry can be relaxed. Given a set S with N elements and a dissimilarity measure δ, the distance matrix $\mathbf{\Delta}$ is a (usually symmetric) $N \times N$ matrix, with all zero entries on the main diagonal, whose (non-negative) generic element at row i and column j is the dissimilarity between i-th and j-th element of S.

Distance metric: A distance metric d is a mathematical function satisfying some given conditions that defines a distance between each pair (x, y) of elements of a set S. A set with a metric is called a metric space. Denoted with $d(x, y)$, the distance between x and y, the conditions to be satisfied, for all $x, y, z \in S$ are: non-negativity ($d(x, y) \geq 0$), identity ($d(x, y) = 0 \Leftrightarrow x = y$), symmetry ($d(x, y) = d(y, x)$), triangle inequality ($d(x, y) \leq d(x, z) + d(y, z)$). Based on this definition, several distance metrics can be defined. The most commonly used are the metrics in the Minkowski class, among which we find the popular Euclidean metric. Given a metric space with N elements, the distance matrix $\mathbf{D}$ is a symmetric $N \times N$ matrix, with all zero entries on the main diagonal, whose (non-negative) generic element at row i and column j is the distance between i-th and j-th element of S.

Machine learning: Machine learning was born as artificial intelligence techniques in the fields of pattern recognition and computational learning theory; it aims at constructing algorithms able to learn from data and make predictions or uncover hidden insights through learning from historical relationships and trends in the data. Machine learning is employed in a range of tasks where defining explicit models for the relationships present in data is difficult or infeasible. Today, machine learning is closely related to (and often overlaps with) computational statistics, which also deal with prediction-making and data mining by means of algorithms, and takes support from mathematical optimization. The philosophy behind algorithmic modeling is discussed by Breiman (2001b) and a number of machine learning algorithms are well addressed in Friedman et al. (2009) and Witten et al. (2016).

Network analysis: Network analysis is concerned with constructing and analyzing graphs representing either symmetric or asymmetric relations between discrete objects. In computer science, it is strictly connected to graph theory, where a network is defined as a graph in which nodes and/or edges have some given attributes.

Pattern: An intelligible regularity discernible in the way in which something happens.

CHAPTER 4

Finding Groups in Data

T O DISCOVER some hidden framework according to which data could be reorganized and categorized is one of the purposes of Data Mining. Indeed, very commonly in science great efforts are devoted to the definition of taxonomies, as it meets the need of the human brain to identify underlying structures able to shorthand ideas by means of stylized facts.

From a philosophical perspective, to decide whether taxonomy is art or science is still an open question.

Is taxonomy art, or science, or both?

Sydney Anderson
Biologist, Curator of the American Museum of Natural History

Data Mining algorithms try to address the issue as scientifically as possible, defining methods and rules aimed to assign units of observation into classes, which are not defined a priori, and are supposed to somehow reflect the structure of the entities that the data represent (Kaufman and Rousseeuw, 1990; Guyon et al., 2009). The unsupervised classification of individual cases into groups whose profile design spontaneously emerges from data is known as Cluster Analysis, a broad subject which includes several techniques, which differ significantly in their notion of what constitutes clusters and how to efficiently find them.

Cluster Analysis in basketball can be used to cluster players, teams, matches, game fractions, etc.

Examples of players' clustering can be found in Alagappan (2012), Bianchi (2016), Bianchi et al. (2017) and Patel (2017), where players are classified into groups based on personal performance, with the aim of defining new roles, as opposed to the traditional five positions that are quickly becoming outdated. This issue will be more deeply discussed in Section 4.4.

Classification of matches is extensively used by Sampaio and Janeira (2003), Ibáñez et al. (2003), Csataljay et al. (2009) and Lorenzo et al. (2010), a number of works all aimed to investigate the discriminatory power of game statistics between winning and losing teams, and to identify the critical performance indicators that most distinguish between winning and losing performances within matches. The authors carry out their analyses on different datasets (namely, coming from the Portuguese Professional Basketball League, the Junior World Basketball Championship, the European Basketball Championship and the Under-16 European Championships) and in all the cases a preliminary Cluster Analysis is performed on matches, in order to classify them into three types such as tight games, balanced games and unbalanced games, according to game final score differences. Similarly, Sampaio et al. (2010b) address the issue of quantifying what a fast pace in a game contributes to point differentials and examining the game statistics that discriminate between fast- and slow-paced games; Cluster Analysis is then used to classify game pace using ball possessions per game quarter.

Game fractions clustering can be found in Metulini et al. (2017a) and Metulini et al. (2017b), where groups of typical game configurations are identified based on the distance among players, recorded by GPS sensors with an average frequency of about 80 Hz. Each cluster corresponds to a specific game configuration and it is possible to assign it to either one of offense or defense. In addition, transitions between clusters are examined, highlighting the main tendencies in moving from one game configuration to another one.

Although out of the scope of this book, it is worth mentioning the wide use of Cluster Analysis for market segmentation, also in the context of sport management. For example, Ross (2007) identifies segments of NBA spectators based upon the brand associations held for a team, but several examples can be found in the literature on the definition of customer profiles based on surveys or social network data.

In this chapter, after a brief recall (citing relevant sources for further reading) of the main Cluster Analysis algorithms (Section 4.1), applications in basketball with three case studies will be examined (Sections 4.2 and 4.3). Finally a focus topic will be discussed, dealing with new basketball role definitions (Section 4.4).

4.1 CLUSTER ANALYSIS

Cluster Analysis is a classification technique aiming at dividing individual cases into groups (clusters) such that the cases in a cluster are very similar (according to a given criterion) to one another and very different from the cases in other clusters.

As mentioned before, Cluster Analysis is unsupervised, and it should not be confused with supervised classification methods, such as discriminant analysis, where the groups are known a priori and the aim of the analysis is to create rules for classifying new observation units into one or an other of the known groups. On the contrary, Cluster Analysis is an exploratory method that aims to recognize the natural groups that appear in the data structure. In some sense, Cluster Analysis is a dimensionality reduction technique, because the high (sometimes huge) number of units observed at the beginning is reduced to a smaller number of groups that are homogeneous, allowing a parsimonious description and an easy interpretation of the data. Results can then be used for several aims, for example to identify outliers or find out some hidden relationships.

Cluster Analysis is used in many fields, from market research to medicine, information retrieval, linguistics, biology, astronomy, social sciences, archaeology, anthropology, geology, insurance, sports, and many others. It finds several applications in basketball analytics, as already described in the introduction of this chapter. For example, it may be interesting to group a large number of players into homogeneous groups according to their performances and then to examine how the wages of players belonging to same cluster differ.

In order to perform a Cluster Analysis, first of all the variables to be included in the analysis must be selected. This choice should be supported by conceptual considerations and supervised by an expert of the field (the basketball expert in the case of basketball analytics). According to the aim of the analysis, the best set of variables should be chosen. Usually, if available, a plurality of variables is included, in such a way that the elimination of one of them or the addition of a new variable does not substantially modify the identified structure of groups. In general, results can be very dependent on the variables included; therefore, the choice of variables remains a very crucial issue. Once variables are selected, in most situations they must be standardized, that is transformed into unitless variables (having zero mean and unit variance), in order to avoid dependence of results on unit measurement and magnitude of the original input variables. Standardization is automatic in the functions `hclustering` and `kclustering` implemented in R package `BasketballAnalyzeR` to perform Cluster Analysis.

Homogeneity within each cluster and the degree of separation among clusters are measured by referring to a distance metric or a dissimilarity measure (already introduced in Chapter 3, Section 3.3). In general, the ordering of pairs of units (from the most similar to the most different) is sensitive to the selected measure of distance or dissimilarity, so an important step in Cluster Analysis is the choice of the most appropriate measure, which in practice should be based on the type of data at hand: for numerical variables, distance-type metrics are appropriate, while for categorical variables dissimilarity measures should be preferred.

Subsequently, the clustering method must be selected. Usually, we distinguish (agglomerative and divisive) hierarchical clustering and non-hierarchical clustering.

In hierarchical methods, the members of inferior-ranking clusters become members of larger, higher-ranking clusters, so there is a hierarchical attribution of units to clusters. In agglomerative clustering, the starting point is given by as many clusters as units. Then, the most similar units are grouped together, and these first clusters are then merged according to their similarities. In the end, all groups are fused into a single cluster. On the contrary, in divisive clustering, all objects are initially together in a single cluster, which is gradually divided into subgroups. In the end, the procedure results into singleton clusters of individual data units. In both agglomerative and divisive clustering, a cluster on a higher level of the hierarchy always encompasses all clusters from a lower level. The assignment of an object to a certain cluster is irreversible, so there is no possibility of reassigning this object to another cluster.

Nonhierarchical methods produce a single partition, which is obtained by optimizing an objective function that describes how well data are grouped in clusters (typically, this function is expressed in terms of within-group homogeneity). The most widespread method of this kind is k-means clustering, which is also probably the most popular method of Cluster Analysis in general (Hennig et al., 2015). The number of clusters must be set a priori.

In both hierarchical and nonhierarchical clustering, a very important step in the analysis is the evaluation of the obtained partition and the identification of the optimal number of clusters. In general, a partition is satisfactory when homogeneity within clusters is rather high and clusters are well separated. There exists a trade-off between the number of clusters and internal homogeneity: a partition with a smaller number of clusters is easier to interpret and more useful in practice, but a price must be paid in terms of lower within-group homogeneity, because units that are more different from each other are aggregated. In the hierarchical clustering, the issue is to identify the optimal partition, that is the optimal number of clusters in the hierarchical series of partitions that occurred. In the nonhierarchical clustering, the optimal number of clusters is usually identified by running the procedure several times, varying the number of groups, and then selecting the most satisfactory solution, based on some criterion.

In some situations, the goodness of fit of the obtained partition is evaluated resorting to the decomposition of the Total Deviance, TD, (introduced in Chapter 2, Section 2.2.6) in Between Deviance, BD, and Within Deviance, WD; $TD = BD + WD$. In detail, if N subjects are divided into k clusters, for a given variable X we have

$$TD = \sum_{h=1}^{k} \sum_{i=1}^{n_h} (x_{ih} - \mu)^2$$

where x_{ih} is the value assumed by the i-th subject of cluster h, μ is the overall mean of X and n_h is the number of subjects belonging to cluster h $(n_1 + \cdots + n_k = N)$,

$$BD = \sum_{h=1}^{k} (\mu_h - \mu)^2 n_h$$

where μ_h is the mean of X for only those subjects who belong to cluster h $(\sum_{h=1}^{k} \mu_h n_h = N\mu)$, and

$$WD = \sum_{h=1}^{k} \sigma_h^2 n_h$$

where σ_h^2 is the variance of X for only those subjects who belong to cluster h. BD gives a measure of the degree of separation among clusters, while WD is a measure of the internal homogeneity of the clusters (the higher WD, the lower the homogeneity). The ratio BD/TD is the Pearson's correlation coefficient η^2 introduced in Chapter 3, Section 3.1.2 and in this context it is called "explained variance." It measures the clusterization quality with respect to variable X because, in practice, it accounts for the mean dependence of X on the proposed clusterization: high BD/TD means high BD, that is good separation among clusters (the difference between the cluster means μ_h and the overall mean μ is high), and, being TD fixed, low WD, that is high within-cluster homogeneity (low values of the cluster variances σ_h^2). Of course, as will be detailed in Section 4.2.1, the needs of a high value of BD/TD and an interpretable (low) number of clusters must be balanced.

The numerous choices that must be made in Cluster Analysis introduce elements of subjectivity in the results. In general, a solution can be considered good when it remains approximately stable as algorithms change because, in this situation, the clustering solution reflects a real grouping structure existing in the multidimensional data. The procedure of evaluating the results of a clustering algorithm is known as cluster validity and several indexes and procedures have been proposed in the literature (see, for example, Milligan, 1996; Milligan and Cooper, 1985; Hennig et al., 2015).

In the following, we describe two clustering methods followed by applications to basketball data: the k-means clustering and agglomerative hierarchical clustering (Sections 4.2 and 4.3, respectively). A detailed description of those methods can be found, for example, in Hennig et al. (2015), which offers an overview of the traditional Cluster Analysis but also provides several further developments.

4.2 K-MEANS CLUSTERING

The widespread k-means clustering is a nonhierarchical algorithm and, in particular, belongs to the so-called partitioning methods. The number k of clusters must be specified in advance and the procedure runs over the following steps:

1. choose k initial cluster centers (also called initial seeds),

2. assign each subject to its nearest cluster, after evaluation of the distance/dissimilarity of that subject to each cluster's center,

3. compute the centroids of the clusters that have been created (centroids are the geometric centers of the clusters and are computed, for each cluster, averaging the p clustering variables over the subjects belonging to that cluster; in other words, the coordinates of a cluster's centroid are computed by averaging the coordinates of the subjects in the group),

4. re-calculate the distances subjects-centroids and re-assign each subject to its nearest cluster (reallocating subjects that are not in the cluster to which they are closest),

5. continue until convergence, that is when the centroids (and, consequently, the clusters' composition) remain relatively stable, according to a certain criterion (stopping rule).

Very often, the k-means clustering is used adopting the Euclidean distance, because this guarantees the algorithm's convergence. A very interesting implication is that, in this situation, the algorithm implicitly aims at identifying the partition of data into k groups that minimize the within-cluster deviance (WD). The problem cannot be solved by considering every possible partition of the N subjects into k groups, and choosing the one with the lowest WD, because complete enumeration of all the possible partitions is not feasible, even with the fastest computer (for example, $N = 100$ subjects can be clustered into $k = 5$ groups into 6.57×10^{67} different ways). This explains the necessity to use an iterative optimization as explained above.

Euclidean distance is appropriate when the p variables in analysis are numerical. There exist some variants of the k-means clustering suitable for categorical data (for example, the k-medoids algorithm, see Kaufman and Rousseeuw, 1990).

k-means clustering can handle larger datasets than hierarchical procedures. However, the final solution can be affected by outliers and it performs poorly in identifying non-convex clusters (for example, U-shaped clusters).

A very important issue to underline is that k-means clustering can be very sensitive to the choice of initial cluster centers. A hierarchical Cluster Analysis may be used to determine a good number k of clusters to set in advance and, also, to compute the cluster centers that may be used as initial seeds in the nonhierarchical procedure. If, as usual, initial seeds are randomly chosen, a possible solution consists of running the algorithm a certain number of times and choosing the best solution according to a given criterion. This solution is the one implemented in the function `kclustering` of the package `BasketballAnalyzeR` as explained in Section 4.2.1.

4.2.1 k-means clustering of NBA teams

The R function `kclustering` can be used to perform k-means Cluster Analysis in a simple and flexible way. It is suited to avoid as much as possible the possibility of obtaining different clusterizations from one running to another, because it runs the algorithm `nruns` times and chooses the best solution according to a maximum explained variance criterion. By default `nruns=10`, but the argument is customizable by the researcher. By way of example, we now define groups of NBA teams based on game variables specifically built to take into account offensive and defensive skills. To this aim, we also resort to Offensive/Defensive Ratings and Dean Oliver's Four Factors (Chapter 2, Section 2.2.1), focusing attention on the first three Factors, that are usually considered the most influential (the approximate weights Dean Oliver assigned to the Four Factor are 40%, 25%, 20%, 15%). The required statistics are computed thanks to the function `fourfactors`

```
> FF <- fourfactors(Tbox,Obox)
```

We also define ratios of the offensive over the defensive statistics, having the meaning of how well the team performed with respect to the opponents, with better performances corresponding to higher ratios:

- the ratio $ORtg/DRtg$ of the Offensive over the Defensive Rating, defined in (2.3) and (2.4)

  ```
  > OD.Rtg <- FF$ORtg/FF$DRtg
  ```

- the ratio of the offensive over the defensive first Factor $eFG\%$ (Table 2.4)

  ```
  > F1.r <- FF$F1.Off/FF$F1.Def
  ```

- the ratio of the defensive over the offensive second Factor $TO\ Ratio$ (Table 2.4)

  ```
  > F2.r <- FF$F2.Def/FF$F2.Off
  ```

In addition, we consider the third Factor, 3-point shots and steals:

- the offensive third Factor $REB\%$ (Table 2.4)

  ```
  > F3.Off <- FF$F3.Off
  ```

- the defensive third Factor $REB\%$ (Table 2.4)

  ```
  > F3.Def <- FF$F3.Def
  ```

- the total number of 3-point shots made ($P3M$)

  ```
  > P3M <- Tbox$P3M
  ```

- the ratio STL_T/STL_O of the team over the opponents' steals.

```
> STL.r <- Tbox$STL/Obox$STL
```

Finally, the data matrix we use for Cluster Analysis is built as

```
> data <- data.frame(OD.Rtg, F1.r, F2.r, F3.Off, F3.Def,
                     P3M, STL.r)
```

The function `kclustering` has to be used in two steps: firstly without specifying a value for the argument `k`, in order to decide the number of clusters, secondly declaring the number of clusters to be defined. In both cases, objects of class `kclustering` are generated and a `plot` method is available for this class, giving different outputs according to whether the argument `k` is specified or not. With the code

```
> set.seed(29)
> kclu1 <- kclustering(data)
> plot(kclu1)
```

we obtain the graph in Figure 4.1, displaying the pattern of the explained variance, the above-mentioned measure of the clusterization quality (the extent to which the individual cases within the clusters are similar to each other and different from those belonging to other clusters) with respect to the number of clusters. In detail, the solid line represents the (average over all the variables) ratio of the Between over the Total Deviance BD/TD, which improves as the number of clusters increases. In general, values higher than 50%, meaning that the clusterization is able to explain more than half of the total variability, may be considered as satisfactory. Nonetheless, we have to balance two contrasting needs: on one hand, we aim to have a high-quality clusterization, on the other hand, we need to obtain as few clusters as possible, complying with the general statistical criterion of parsimony and to facilitate interpretation. To decide the optimal number of clusters, we consider the dotted line, representing the percentage increase of the ratio BD/TD moving from a $(k-1)$-cluster to a k-cluster solution. We should identify a threshold below, which the improvement obtained thanks to an additional cluster, is too low to justify the higher complexity generated by the additional cluster itself. In general, we may identify this threshold on an elbow of the dotted line, or considering as worthwhile increments greater than, say, about 5-10%. In this case, also in view of the low number of subjects (30 teams), the optimal number of clusters can be identified as 5 clusters, with a total clusterization quality of $BD/BT = 59.13\%$.

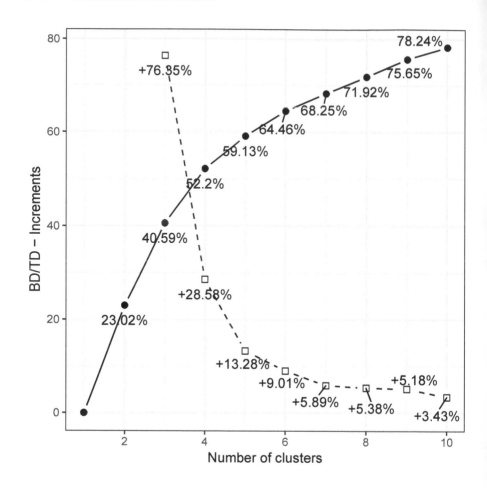

Figure 4.1 Pattern of the clusterization quality with respect to the number of clusters (*y*-axis: percentage increase in the ratio of the Between over the Total Deviance, *BD/BT*).

The code lines to carry out the second step of the analysis are then

```
> set.seed(29)
> kclu2 <- kclustering(data, labels=Tbox$Team, k=5)
> plot(kclu2)
```

The R object `kclu2` is a list of data frames containing the subjects' cluster identifiers, the clusters' composition, the clusters' profiles (*i.e.*, the average of the variables within the clusters) and, for each cluster, the Cluster Heterogenity Index (*CHI*), an average measure of the variability within the cluster, which should be as low as possible. In detail, the index CHI for cluster h is computed as

$$CHI_h = \frac{\sum_{j=1}^{p} \sigma_{jh}^2}{p}$$

where σ_{jh}^2 is the variance of the j-th variable X_j within cluster h. More specifically, $CHI = 1$ denotes that the cluster has, on average, the same variability of the entire dataset, so that the clusterization (at least with reference to that cluster) is undermined. In general, we may consider as satisfactory values lower than 50%, but when the number of cases to be classified is high, this threshold has to be raised and it is common to obtain some highly heterogeneous clusters among the others. The obtained clusters are summarized in Table 4.1. It is worth pointing out that, due to the iterative nature of the k-means algorithm, the order of clusters changes from one running to another, so the interpretation must be adapted correspondingly (in order to let you obtain the same order of our analysis, we have added set.seed(29) just before running the k-means algorithm). In this case, the plot function generates the graph in Figure 4.2, where a radial plot (see Chapter 2, Section 2.2.3) of the average profiles of the clusters is visualized, together with the

TABLE 4.1 Clusters' composition. Note: the order of clusters changes from one running to another; to obtain the same order, use set.seed(29).

Cluster 1
Atlanta Hawks, Brooklyn Nets, Memphis Grizzlies, New York Knicks, Orlando Magic, Phoenix Suns
Cluster 2
Charlotte Hornets, Chicago Bulls, Dallas Mavericks, Detroit Pistons, Sacramento Kings
Cluster 3
Denver Nuggets, LA Clippers, Los Angeles Lakers Philadelphia 76ers, Portland Trail Blazers, San Antonio Spurs, Washington Wizards
Cluster 4
Indiana Pacers, Milwaukee Bucks, Minnesota Timberwolves, Oklahoma City Thunder
Cluster 5
Boston Celtics, Cleveland Cavaliers, Golden State Warriors Houston Rockets, Miami Heat, New Orleans Pelicans, Toronto Raptors, Utah Jazz

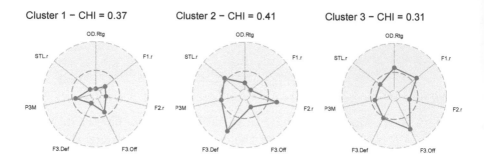

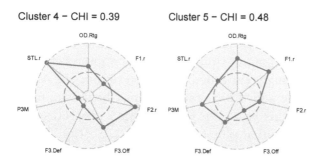

Figure 4.2 Radial plots of the average profiles of NBA teams' clusters. CHI = Cluster Heterogenity Index; OD.Rtg = ratio $ORtg/DRtg$ of the Offensive over the Defensive Rating; F1.r = ratio of the offensive over the defensive first Factor $eFG\%$ (Table 2.4); F2.r = ratio of the defensive over the offensive second Factor $TO\ Ratio$ (Table 2.4); F3.Off = offensive third Factor $REB\%$ (Table 2.4); F3.Def = defensive third Factor $REB\%$ (Table 2.4); STL.r = ratio STL_T/STL_O of the team's over the opponents' steals.

CHI index. In interpreting the radial plots, we must consider that the variables used for Cluster Analysis are previously standardized, so that they are all expressed on the same scale, and the blue dashed line denotes the overall (zero) mean.

Observing the results, we can draw some interesting remarks. First, moving from cluster 1 to 5, we also tend to move from the bottom to the top ranked teams, so that we can identify the game features corresponding to different achievements in the championship. In detail, cluster 1 is composed of teams generally bottom ranked, with values lower than the average in all the considered variables. In cluster 2 we find medium-bottom ranked teams, which exhibit, on average, low offensive abilities compared to their opponents, but good defensive skills, as denoted by high values for the second Factor ratio and the defensive third Factor. Cluster 3 contains teams that obtained different final ranking position, including Philadelphia 76ers and Portland Trail Blazers, which both ranked 3rd in their Conference. The average profile of this cluster denotes fairly good offensive statistics and values little below the average in defensive ones. Cluster 4 has very interesting connotations: fairly good as for the Offensive over Defensive Rating ratio, but lower than the average in the first Factor ratio; remarkably high values for the second Factor ratio, the offensive third Factor and the steals ratio, but remarkably low values in the defensive third Factor and the 3-point shots. In summary, teams in this cluster seem to exhibit high athleticism (as reflected by good performance in turnovers, steals, offensive rebounds), but, overall, have a poorly effective way of playing (low field goal percentages, low number of 3-point shots made). It is worth noting that the Milwaukee Bucks and Oklahoma City Thunder are the two top-ranked teams as for scored points inequality, measured with the Gini coefficient (Chapter 2, Section 2.2.6). All the teams in this cluster qualified for the Playoffs. The last cluster is composed of top ranked teams, with the exception of the Miami Heat, New Orleans Pelicans and Utah Jazz, that ranked 5th and 6th in their Conference. The Boston Celtics, Cleveland Cavaliers, Golden State Warriors and Houston Rockets are the four Conference finalists. The average profile in this cluster denotes high values in the Offensive over Defensive Rating ratio, in the first Factor ratio and in the 3-point shots; on the average or little above for the defensive skills denoted by the second Factor ratio, the defensive third factor and the steals; remarkably low the value in the offensive third Factor. This seems to indicate that a good efficacy of possessions, high field goal percentages and a high exploitation of 3-point shots results in good global achievements for the team, even in presence of no outstanding values for statistics related to turnovers, steals, and rebounds. Actually, this cluster is the most heterogeneous ($CHI = 0.48$, a value just about acceptable, considered the low number of cases in this Cluster Analysis). For this reason, it might be useful to inspect the single teams' profiles and consider that the above remarks, based on the average, have to be taken with a grain of salt.

Further analyses of the obtained clusters can be carried out by resorting to the basic functions described in Chapter 2. For example, we could deepen the investigation of the different achievements of the teams belonging to different clusters, in the considered championship. Firstly, we evaluate how many teams of each cluster qualified or not for the Playoffs (kclu2.PO) and compute the average number of wins for the teams belonging to each cluster (kclu2.W)

```
> kclu2.PO <- table(kclu2$Subjects$Cluster, Tadd$Playoff)
> kclu2.W <- tapply(Tbox$W, kclu2$Subjects$Cluster, mean)
```

Secondly, using the function barline, we generate the bar-line plot in the top panel of Figure 4.3, showing that clusters 1 and 2 are composed only of teams that did not qualify for the Playoffs, cluster 3 of some teams that qualified and some others that did not qualify, clusters 4 and 5 only comprise qualified teams. As expected, the number of average wins increases from cluster 1 to 5.

```
> Xbar <- data.frame(cluster=c(1:5), N=kclu2.PO[,1],
                Y=kclu2.PO[,2], W=kclu2.W)
> barline(data=Xbar, id="cluster", bars=c("N","Y"),
        labels.bars=c("Playoff: NO","Playoff: YES"),
        line="W", label.line="average wins",
        decreasing=FALSE)
```

In addition, using the function bubbleplot, we may generate a bubble plot of the teams, displaying the points scored by the team (x-axis) and by its opponents (y-axis), the number of wins (bubble size) and the cluster to which the team has been assigned (bubble color)

```
> cluster <- as.factor(kclu2$Subjects$Cluster)
> Xbubble <- data.frame(Team=Tbox$Team, PTS=Tbox$PTS,
                    PTS.Opp=Obox$PTS, cluster,
                    W=Tbox$W)
> labs <- c("PTS", "PTS.Opp", "cluster", "Wins")
> bubbleplot(Xbubble, id="Team", x="PTS", y="PTS.Opp",
            col="cluster", size="W", labels=labs)
```

The outcome is plotted in the bottom panel of Figure 4.3, showing that the teams of the top-performing cluster 5 tend to be spread in all the quadrants (except the top-left one). This means that their similar way of playing (with respect to the variables considered for the Cluster Analysis) can lead to different kinds of achievements, characterized by both a low and high number of points scored and suffered. The Boston Celtics, Miami Heat and Utah Jazz score below the average, but they also suffer few points from the opponents. Going back to the graph in the top

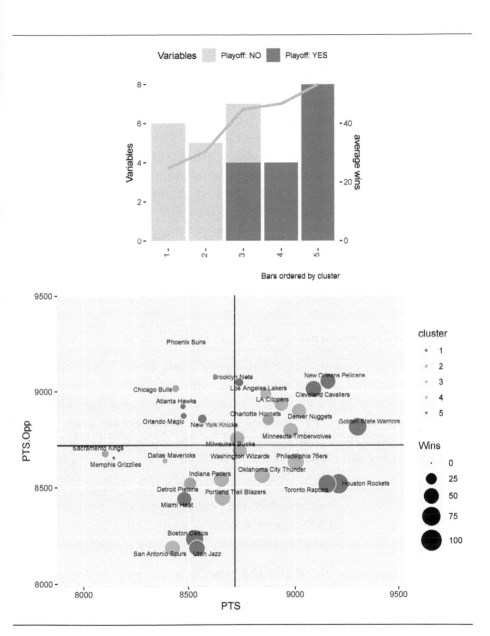

Figure 4.3 Top: Bar-line plot of the achievements of teams belonging to
the 5 clusters: qualification for Playoffs (yes or no) for the bars and
average number of wins for the line. Bottom: Bubble plot of the teams
for points scored by the team (x-axis) and by its opponents (y-axis),
number of wins and cluster.

panel of Figure 2.2, we recall that these three teams tend to play at a low pace, which can justify the low number of scored and suffered points. The best positioned and top-ranked teams are the Toronto Raptors, Houston Rockets (that score more points and suffer less than the average) and Golden State Warriors (that has the highest value of scored points and its opponents scored a little above the average). On the other hand, the New Orleans Pelicans and Cleveland Cavaliers score and suffer more points than the average. Similarly, also the teams of clusters 3 and 4 tend to be scattered in the same three quadrants just analyzed, while the teams of clusters 1 and 2 can be found in the top-left and bottom-left quadrants. The Charlotte Hornets and Detroit Pistons deserve a comment of their own, as they are positioned amongst the teams of clusters 3, 4 and 5. Evidently, despite the fact that they belong to cluster 2, their game features considered in the Cluster Analysis are more similar to those of the teams in cluster 2; they have better achievements than the other teams of their cluster.

4.2.2 k-means clustering of Golden State Warriors' shots

Another interesting example of Cluster Analysis can be obtained using the shots as subjects, with the aim of grouping shots that are similar to each other. From the play-by-play dataset PbP (obtained from PbP.BDB thanks to the function PbPmanipulation as described on page 23) we extract the 6979 field goals attempted by the players of Golden State Warriors during the regular season 2017/2018

```
> shots <- subset(PbP,
                !is.na(PbP$shot_distance) &
                PbP$team=="GSW")
> shots <- dplyr::mutate_if(shots, is.factor, droplevels)
```

where the last code line drops the unused levels from factors in the data frame shots. We cluster the shots with respect to points scored, shot distance, time in the quarter and play length, following the two-step procedure already described for the example in Section 4.2.1 (in this case, considered the graph in Figure 4.4, we decide to perform a 6-cluster clusterization, with an overall quality of 58.53%). As in the previous example, the order of clusters can change from one running to another, so we suggest to set a seed (to 1) in order to obtain the same solution we are going to describe hereafter.

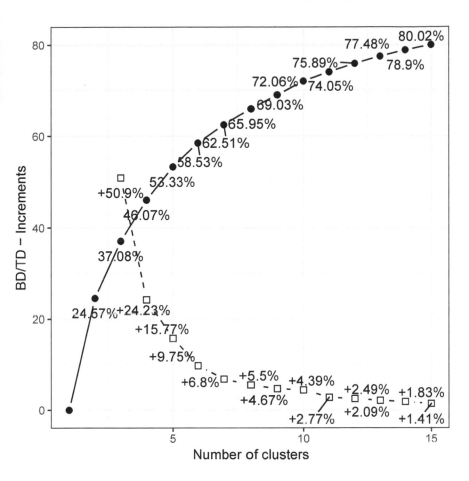

Figure 4.4 Pattern of the clusterization quality with respect to the number of clusters (y-axis: percentage increase in the ratio of the Between over the Total Deviance, BD/BT).

```
> attach(shots)
> data <- data.frame(PTS=points, DIST=shot_distance,
                     TIMEQ=periodTime, PL=playlength)
> detach(shots)
> set.seed(1)
> kclu1 <- kclustering(data, algorithm="MacQueen",
                       nclumax=15, iter.max=500)
> plot(kclu1)
```

The function `kclustering` admits to set some parameters related to the k-means method and the specific algorithm to be used. In addition to the above-mentioned `nruns`, we can tune the maximum value of k in the first step analysis (`nclumax`) and the maximum number of allowed iterations (`iter.max`). As for the algorithms, all the options of the general R function `kmeans` are available, namely the algorithms proposed by Hartigan and Wong (Hartigan and Wong, 1979) (default), MacQueen (MacQueen, 1967), Lloyd (Lloyd, 1982) and Forgy (Forgy, 1965). In this example, we tuned the parameters `nclumax` and `iter.max` and we used MacQueen's algorithm.

```
> set.seed(1)
> kclu2 <- kclustering(data, algorithm="MacQueen",
                       iter.max=500, k=6)
> plot(kclu2)
```

The radial plots in Figure 4.5 show that the clusters are moderately homogeneous ($CHI \leq 0.5$) and have well-defined profiles, which can be easily interpreted in terms of the variables selected for the clusterization. A better representation can be obtained by resorting to shot charts, using the function `shotchart`. We first add the cluster identifiers to the data frame `shot` and we adjust the shot coordinates as described in Chapter 2, Section 2.2.8.

```
> cluster <- as.factor(kclu2$Subjects$Cluster)
> shots <- data.frame(shots, cluster)
> shots$xx <- shots$original_x/10
> shots$yy <- shots$original_y/10 - 41.75
```

Subsequently, for each cluster we draw two shot plots: one displaying the shots according to whether they are missed or made, and one with sectors colored according to the time in the quarter and annotated with scoring percentages. To do that, we use the following code lines with a `for` loop

```
> no.clu <- 6
> p1 <- p2 <- vector(no.clu, mode="list")
> for (k in 1:no.clu) {
      shots.k <- subset(shots,cluster==k)
      p1[[k]] <- shotchart(data=shots.k, x="xx", y="yy",
                           z="result", type=NULL,
                           scatter = TRUE,
                           drop.levels=FALSE)
```

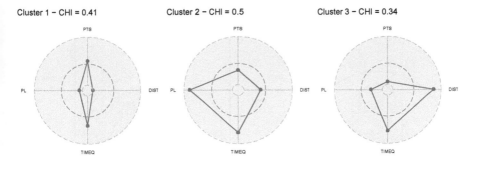

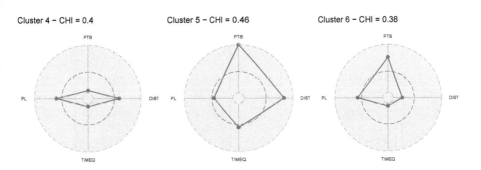

Figure 4.5 Radial plots of the average profiles of shot clusters. $CHI =$ Cluster Heterogeneity Index; DIST = shot distance; TIMEQ = time in the quarter; PL = play length.

```
p2[[k]] <- shotchart(data=shots.k, x="xx", y="yy",
                     z="periodTime",
                     col.limits=c(0,720),
                     result="result", num.sect=5,
                     type="sectors", scatter=FALSE)
}
```

The shot plots are arranged in a grid and displayed in Figures 4.6 and 4.7:

```
> library(gridExtra)
> grid.arrange(grobs=p1, nrow=3)
> grid.arrange(grobs=p2, nrow=3)
```

The shot plots of Figure 4.6 clearly show that 4 clusters out of 6 are homogeneously composed by shots all made or missed. Only clusters 1 and 2 contain made and missed shots mixed, with a prevalence of made

Figure 4.6 Shot plots of the clusters (by row): shots' scatter and missed/made color coding.

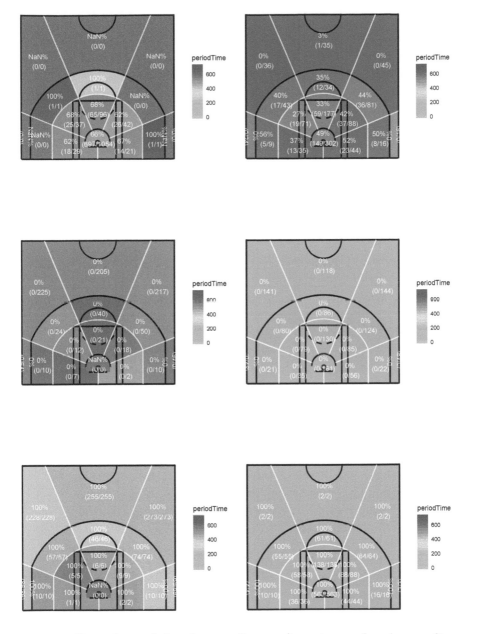

Figure 4.7 Shot plots of the clusters (by row): sectors colored according to the time in the quarter (periodTime) and annotated with scoring percentages.

and missed in clusters 1 and 2, respectively. Clusters 1 and 6 tend to contain close- and mid-range shots, clusters 3 and 5 long-range shots, and clusters 2 and 3 all kinds of shots from what concerns the distance from the basket. The additional information of Figure 4.7 is related to the time in the quarter: the shots in clusters 1, 2 and 3 are attempted on average in the second part of the quarter, while those in clusters 4 and 6 are attempted in the first part. Since the majority of missed 3-point shots are concentrated in clusters 3 and 4, we conclude that they tend to occur respectively late and early in the quarter. Adding the information about the play length deduced by the radial plots (Figure 4.5), we also observe that the missed shots of cluster 3 tend also to be attempted in the early part of the play.

At a second stage, we may investigate the inequality of clusters with respect to the players attempting the shots. A cluster with perfect equality would be composed by shots equally distributed among all the players, while high levels of inequality would mean that a big fraction of shots has been attempted by a small fraction of players. To do that, we first obtain a cross-table of the shots attempted by the players in the 6 clusters (`shots.pl`, Table 4.2), then we arrange it as a data frame (`Xineq`)

TABLE 4.2 Cross-table of the shots attempted by the players in the 6 clusters (C1, ..., C6).

Player	C1	C2	C3	C4	C5	C6
Andre Iguodala	81	64	45	55	38	37
Chris Boucher	0	1	0	0	0	0
Damian Jones	7	12	1	0	0	2
David West	58	24	9	117	16	154
Draymond Green	133	63	73	167	78	105
JaVale McGee	95	54	8	18	3	41
Jordan Bell	104	38	5	13	2	23
Kevin Durant	194	229	201	198	249	150
Kevon Looney	83	61	6	15	3	24
Klay Thompson	80	93	173	348	273	212
Nick Young	24	69	129	92	128	46
Omri Casspi	72	49	6	23	10	47
Patrick McCaw	48	59	40	36	22	21
Quinn Cook	46	45	47	36	58	20
Shaun Livingston	66	51	9	99	5	113
Stephen Curry	111	115	224	106	242	66
Zaza Pachulia	80	32	2	56	4	89

```
> shots.pl <- table(shots$player, shots$cluster)
> Xineq <- as.data.frame.matrix(shots.pl)
```

Then, for each cluster, we plot the Lorenz curve and compute the Gini coefficient by means of the function inequality. To do that, we use the following code lines with a for loop

```
> no.clu <- 6
> p <- vector(no.clu, mode="list")
> for (k in 1:no.clu) {
        ineqC <- inequality(Xineq[,k], npl=nrow(Xineq))
        title <- paste("Cluster", k)
        p[[k]] <- plot(ineqC, title=title)
        }
```

In the end, we arrange all the plots in a unique frame

```
> library(gridExtra)
> grid.arrange(grobs=p, nrow=3)
```

The graphs in Figure 4.8 show that the clusters with the highest levels of inequality are those where the majority of 3-point shots are attempted (clusters 3 and 5), both made and missed. On the other hand, the lowest levels of inequality are in clusters 1 and 2, containing for the main part close and mid-range shots, both made and missed, and a small part of 3-point shots, all missed. This begs the question, how are the players' shots distributed among the clusters?

To answer this further question, we resort again to a bar-line plot, by means of the function barline. We obtain the frequency distribution of each player's shots in the clusters (shots.perc, Table 4.3) and we arrange it as a data frame, with the additional variable of the total number of shots attempted by each player (Xbar)

```
> shots.perc <- shots.pl/rowSums(shots.pl)
> Xbar <- data.frame(player=rownames(shots.pl),
                    rbind(shots.perc),
                    FGA=rowSums(shots.pl))
> labclusters <- c("Cluster 1","Cluster 2","Cluster 3",
                    "Cluster 4","Cluster 5","Cluster 6")
```

Finally, we draw the bar-line plot using the shots attempted in the clusters for the bars and the total attempted shots for the line

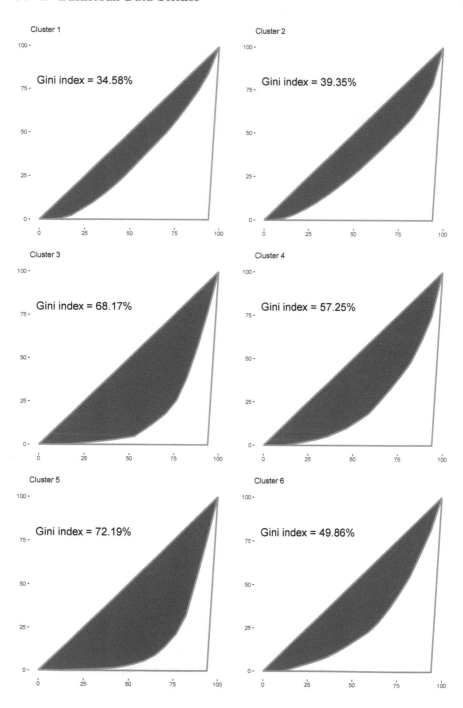

Figure 4.8 Inequality plots of the clusters with respect to the players attempting the shots.

TABLE 4.3 Frequency distribution (%) of players' shots in the 6 clusters (C1,...,C6).

Player	C1	C2	C3	C4	C5	C6
Andre Iguodala	25.31	20	14.06	17.19	11.88	11.56
Chris Boucher	0	100	0	0	0	0
Damian Jones	31.82	54.55	4.55	0	0	9.09
David West	15.34	6.35	2.38	30.95	4.23	40.74
Draymond Green	21.49	10.18	11.79	26.98	12.60	16.96
JaVale McGee	43.38	24.66	3.65	8.22	1.37	18.72
Jordan Bell	56.22	20.54	2.70	7.03	1.08	12.43
Kevin Durant	15.89	18.76	16.46	16.22	20.39	12.29
Kevon Looney	43.23	31.77	3.13	7.81	1.56	12.50
Klay Thompson	6.79	7.89	14.67	29.52	23.16	17.98
Nick Young	4.92	14.14	26.43	18.85	26.23	9.43
Omri Casspi	34.78	23.67	2.90	11.11	4.83	22.71
Patrick McCaw	21.24	26.11	17.70	15.93	9.73	9.29
Quinn Cook	18.25	17.86	18.65	14.29	23.02	7.94
Shaun Livingston	19.24	14.87	2.62	28.86	1.46	32.94
Stephen Curry	12.85	13.31	25.93	12.27	28.01	7.64
Zaza Pachulia	30.42	12.17	0.76	21.29	1.52	33.84

```
> barline(data=Xbar, id="player", line="FGA",
        bars=c("X1","X2","X3","X4","X5","X6"),
        order.by="FGA", label.line="Field goals attempted",
        labels.bars=labclusters)
```

The graph displayed in Figure 4.9, read side-by-side with Table 4.3, informs us about the shooting patterns of the players with respect to the clusters. The players with the highest number of attempted shots (Kevin Durant, Klay Thompson, Stephen Curry, Draymond Green and Nick Young) tend to shoot quite balancedly in the 6 clusters, albeit with some differences (more than half of Klay Thompson's shots are in clusters 4 and 5, while Stephen Curry's and Nick Young's shots are for the main part in clusters 3 and 5). The players with an intermediate number of shots (from David West to Jordan Bell), instead, tend to be more concentrated in some specific cluster. For example, more than 85% of Zaza Pachulia's and David West's shots are in clusters 1, 4 and 6, while Jordan Bell's shots are highly concentrated in clusters 1 and 2.

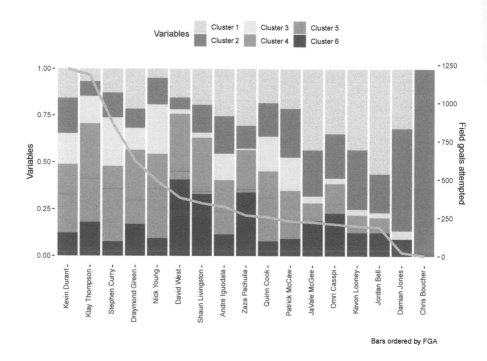

Figure 4.9 Bar-line plot: shots attempted by the players in the clusters.

4.3 AGGLOMERATIVE HIERARCHICAL CLUSTERING

Agglomerative hierarchical clustering aims at building a hierarchy of clusters. The strategy of the agglomerative hierarchical clustering starts with all observation units in their own group; then, the two groups that have the smallest distance or dissimilarity are repeatedly merged, until there is only one cluster, containing all the observation units together.

In detail, if we want to cluster N subjects based on p variables (for example, N players based on p performance indicators), we start with N singleton clusters (each cluster contains only one subject) and a symmetric matrix of distances $\mathbf{D}^p$ or dissimilarities Δ (see also Chapter 3, where $\mathbf{D}^p$ and Δ have been introduced in Section 3.3). Then, the steps in the agglomerative hierarchical clustering algorithm are the following (Johnson and Wichern, 2013):

1. search the distance matrix for the nearest pair of clusters, that is those whose distance or dissimilarity is the lowest. Let the distance between the nearest clusters I and J be d_{IJ},

2. merge clusters I and J and label this new cluster (IJ). Update the values in the distance matrix by deleting the rows and columns corresponding to clusters I and J and adding new rows and columns that collect the distances between cluster (IJ) and the remaining clusters,

3. repeat Steps 1 and 2 for $N-1$ times.

In the end, all subjects belong to a single cluster. At each step, it is useful to record the identity of clusters that are merged and the levels (distances or dissimilarities) at which the clusters are merged.

The results of hierarchical clustering are usually graphically presented in a dendrogram or cluster tree, as the one in Figure 4.12, a tree diagram used to illustrate the sequence of cluster fusions and the distance/dissimilarity at which each agglomeration took place. A careful inspection of the dendrogram is needed to identify the best partition, which should be verified using internal criteria, that is evaluating the resulting data partition using information obtained from the clustering algorithm, or even external criteria, using information obtained from outside the clustering procedure (which, however, is not often available when, as usual, the true cluster structure is not known a priori).

Results from a hierarchical clustering procedure depend on the choice of the distance (or dissimilarity) measure used in the procedure as well as the method defining the distance (or dissimilarity) between two groups. The most common methods, listed below, compute the distance between two clusters as:

- *Single linkage (nearest neighbor)*: the shortest distance between any two members in the two clusters,

- *Complete linkage (furthest neighbor)*: the longest distance between any two members in the two clusters,

- *Average linkage*: the average distance between all pairs of the two clusters' members.

Other widespread methods, using the original data matrix besides the distance (or dissimilarity) matrix are:

- *Centroid method*: the distance between two clusters is given by the distance between the two centroids, which are the geometric centers of the clusters, as explained before (Section 4.2),

- *Ward minimum deviance method*: The agglomeration is not obtained by merging the two most similar subjects successively, but clusters are created by choosing the merging of subjects that increases the overall within-cluster deviance (WD) to the smallest possible degree.

Each method has its own characteristics and allows to identify clusters with different peculiarities: for example, the Ward method aims at finding compact, spherical clusters; single linkage allows to detect clusters that have curvy shapes (instead of spherical or elliptical shapes) and is robust to outliers; complete linkage is more sensitive to outliers and tends to produce clusters of the same size and shape. In Section 4.3.1, an agglomerative hierarchical clustering is performed, using the `hclustering` function available in the R package `BasketballAnalyzeR`, which leans on the `hclust` function in R. All the above-listed methods are there available (in particular, there are two versions of the Ward method, differing on the way distances/dissimilarities are treated before cluster updating).

4.3.1   Hierarchical clustering of NBA players

The R function `hclustering` can be used to perform agglomerative hierarchical clustering, allowing to use all the linkage methods supported by the general `hclust` function, on which it is based (default = Ward method), as mentioned in Section 4.3. Pairwise dissimilarity is based on the Euclidean distance. We show an example of hierarchical clustering of a selection of NBA players based on the main game statistics of the boxscores

```
> attach(Pbox)
> data <- data.frame(PTS, P3M, REB=OREB+DREB,
                     AST, TOV, STL, BLK, PF)
> detach(Pbox)
```

We carry out the analysis on only those players who have played at least 1500 minutes

```
> data <- subset(data, Pbox$MIN>=1500)
> ID <- Pbox$Player[Pbox$MIN>=1500]
```

Analogously to `kclustering`, the function `hclustering` has to be used in two consecutive steps: firstly without specifying a value for the argument `k`, in order to generate the plot of the ratio BD/TD and decide the number of clusters, secondly declaring the number of clusters to be defined (argument k). In both cases, an R object of class `hclustering` is generated. A plot method is available for this class, allowing a lot of graphical options for the dendrogram. So, in the first step, with the code line

```
> hclu1 <- hclustering(data)
> plot(hclu1)
```

we obtain the graph in Figure 4.10. In this case, considered both the quite high number of subjects (183 players) and the percentage increase

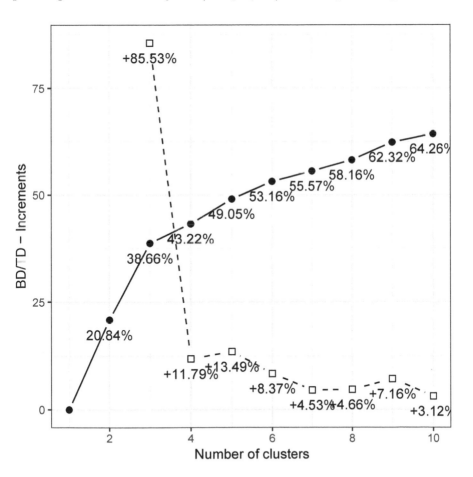

Figure 4.10   Pattern of the clusterization quality with respect to the number of clusters (y-axis: percentage increase in the ratio of the Between over the Total Deviance, BD/BT).

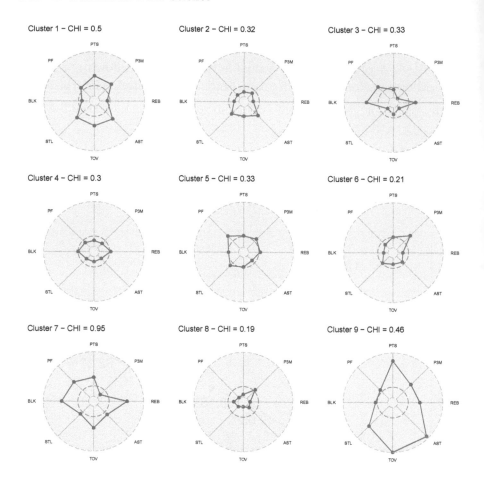

Figure 4.11 Radial plots of the average profiles of NBA players' clusters. CHI = Cluster Heterogeneity Index.

of 7.16% in the clusterization quality when moving from an 8-cluster to a 9-cluster solution, we choose to carry out the analysis with 9 clusters, with an overall clusterization quality of 62.32%. We also display the radial plots of the cluster profiles (Figure 4.11) through the following code lines

```
> hclu2 <- hclustering(data, labels=ID, k=9)
> plot(hclu2, profiles=TRUE)
```

The radial plots allow us to get an idea of the average playing style of the players belonging to different clusters. The CHI index warns us about a low representativity of the profiles for cluster 7 and, to a lesser extent, clusters 1 and 9. The other clusters exhibit very satisfactory values, so we can say that, for example, cluster 3 contains players characterized by rebounds (defensive and offensive), blocks and personal fouls above the average. Cluster 6 is distinguished by the high number of 3-point shots made.

Finally, we produce the dendrogram of the agglomeration, which shows the sequence of cluster fusions and the distance at which each fusion took place (Figure 4.12), with rectangles (rect=TRUE) drawn around colored (colored.branches=TRUE) branches highlighting the corresponding cluster

```
> plot(hclu2, rect=TRUE, colored.branches=TRUE,
      cex.labels=0.5)
```

Similarities among players can be evaluated based on the height at which branches merge. For example, the players of cluster 7 are merged with long branches, which is consistent with the high values of the CHI index for that cluster. Looking at the dendrogram, we see that even a further division of the cluster in two sub-clusters would not completely explain the heterogeneity among the players in that part of the cluster tree. The same can be said for cluster 9 that contains just 3 players. Cluster 1 deserves a different comment: it could be divided into 2, 3 or even 4 sub-clusters with a significant shortening of the branches' merging height.

A complementary visualization of the cluster profiles can be obtained using the variability function, described in Chapter 2, Section 2.2.6. With respect to radial plots, it has the merit of clearly highlighting variability within the clusters. By way of example, we illustrate its use with the clusterization in hclu2. We first compose the data frame X with the cluster identifier and the (standardized) data used for the clustering, and add the minutes played by the analyzed players to be used as additional information for the graph

```
> Pbox.subset <- subset(Pbox, MIN>=1500)
> MIN <- Pbox.subset$MIN
> X <- data.frame(hclu2$Subjects, scale(data), MIN)
```

Afterwards, for each cluster, we plot the variability diagram, using the minutes played to set the size of the bubbles. Note that we set weight = FALSE because in the clustering framework we do not compute weighted variability measures and VC=FALSE because the variables are not strictly positive, so the variation coefficient cannot be used. Again, we use a for loop and then arrange the graphs in a grid

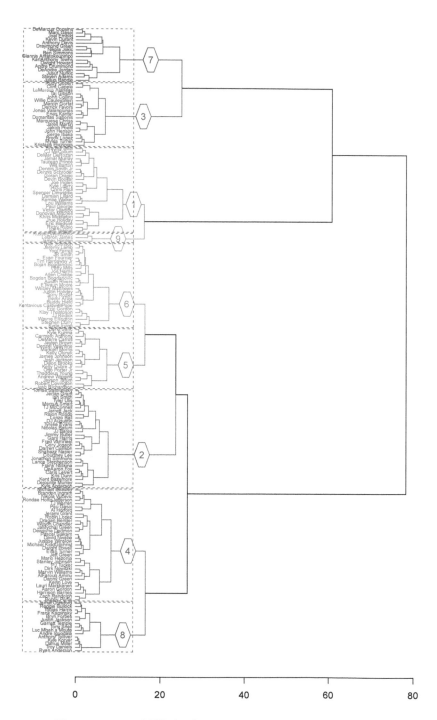

Figure 4.12 Cluster tree of NBA players.

```
> dvar <- c("PTS","P3M","REB","AST",
            "TOV", "STL","BLK","PF")
> svar <- "MIN"
> yRange <- range(X[,dvar])
> sizeRange <- c(1500, 3300)
> no.clu <- 9
> p <- vector(no.clu, mode="list")
> for (k in 1:no.clu) {
      XC <- subset(X, Cluster==k)
      vrb <- variability(XC[,3:11], data.var=dvar,
                          size.var=svar, weight=FALSE,
                          VC=FALSE)
      title <- paste("Cluster", k)
      p[[k]] <- plot(vrb, size.lim=sizeRange, ylim=yRange,
                     title=title, leg.pos=c(0,1),
                     leg.just=c(-0.5,0),
                     leg.box="vertical",
                     leg.brk=seq(1500,3000,500),
                     leg.title.pos="left", leg.nrow=1,
                     max.circle=7)
      }
> library(gridExtra)
> grid.arrange(grobs=p, ncol=3)
```

The graphs in Figure 4.13 show the differences between the clusters in terms of the average values of the variables used for the clusterization. They provide the same information as the radial plots of Figure 4.11, but here we have the additional visualization of the subjects' variability within clusters and the minutes played. It is interesting to note, for example, the high dispersion of all the variables in cluster 7, which exhibited a high value of the heterogeneity index $(CHI = 0.95)$. Another important point concerns cluster 1, whose quite high overall heterogeneity $(CHI = 0.5)$ seems to be due to the high variability of only a limited number of variables (namely, STL, PF, $P3M$, PTS). On the other hand, there are clusters with a satisfactory heterogeneity index where we can find anyway one or two variables with high variability: for example, variable BLK in cluster 3 $(CHI = 0.33)$. Finally, the additional information about the minutes played helps give a better understanding of the analyzed data. We note for example that cluster 8, characterized by values below the average in all the variables, is composed of players

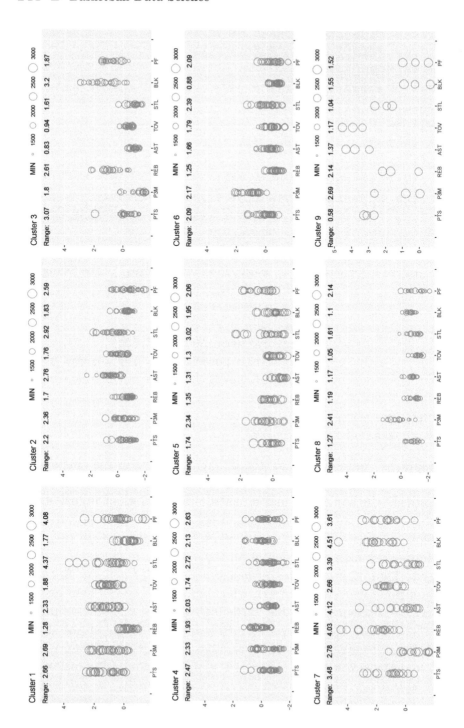

Figure 4.13 Variability diagrams within the clusters.

with small bubbles (anyway, we recall that players in the analysis played at least 1500 minutes), which could in part explain their low stats.

4.4 FOCUS: NEW ROLES IN BASKETBALL

The five positions normally employed in basketball are the point guard (PG), the shooting guard (SG), the small forward (SF), the power forward (PF), and the center (C)[1]. But the way of playing basketball has profoundly changed since James Naismith invented the game for his students. One of the most disrupting change in the game was the introduction of the 3-point line in the early 1980s, which deeply changed the way of playing the game all over the world (Sanders, 1981; Butts, 1986; Lynch, 1987). The availability of a new kind of shot, rewarded with one extra point, obviously encouraged to shoot from a higher distance, thus increasing the importance of long shooting skills and changing the way of playing far from the basket. Overall, the evolution of the game since its birth in 1891 led to a consequent evolution of players, who became more athletic, more skilled, and developed different ways to play the game, so that today there is general consensus about the fact that it may be difficult in some cases to assign a player to one of the 5 positions that have been traditionally used in basketball.

To face changing times is a hot issue in other sports as well. In history, there are widely acknowledged cases of outstanding players, who definitely changed the way of thinking to a specific position, or who defined a new role related to the way they were playing the game. In American football, the case of the *mobile quarterback* is addressed in Monson (2012). In soccer, Corsello (2016) describes why Manuel Neuer, one of the best goalkeepers nowadays, plays like no one goalkeeper before him, and Mahmood (2015) explains how *the Makelele role* redefined English football. Getting back to basketball, Flannery (2016) discusses how Draymond Green is redesigning the concept of NBA basketball superstar.

From an analytic point of view, the first attempt to formalize a quantitative approach for characterizing the basketball players' roles is presented in Alagappan (2012), who exploits Topological Data Analysis

---

[1]For the record, we must mention that historically, only three positions were used (two guards, two forwards, and one center) based on where the player tends to play on the court. Furthermore, some nonstandard positions were recognized in the 1950s, such as the point forward (a hybrid PG/SF), the swingman (a hybrid SG/SF), the big (a hybrid C/PF), and the stretch four (a PF with the shooting pattern of a typical SG).

(TDA, see Carlsson, 2009) to create a sort of map, showing differences between players' statistical profiles based on game statistics, with the aim of redefining basketball roles. The result of an analysis on NBA players' data is a set of 13 clusters, corresponding to 13 new positions, based on the three typical roles of guard, forward and center:

1. **Offensive Ball-Handler:** a cluster of players who handle the ball and specialize in points, free throws and shots attempted, but perform below average in steals and blocks.

2. **Defensive Ball-Handler:** a cluster of defense-minded players who handle the ball and specialize in assists and steals, but are not so good or just on average for points, free throws and shots.

3. **Combo Ball-Handler:** a cluster of players devoted to both offense and defense but not remarkably good in either category.

4. **Shooting Ball-Handler:** a cluster of players with special skills for scoring, characterized by appreciably above average field goal attempts and points.

5. **Role-Playing Ball-Handler:** a cluster of players with a low statistical impact on the game, who often play a few minutes.

6. **3-Point Rebounder:** a cluster of big ball-handlers, above average in rebounds and 3-point shots.

7. **Scoring Rebounder:** a cluster of players above average in rebounds and with a particular attitude for scoring.

8. **Paint Protector:** a cluster of players above average in rebounds, blocks and fouls, but usually below average in scored points.

9. **Scoring Paint Protector:** a cluster of outstanding players both on offense and defense, with special skills in scoring, rebounding and blocking shots.

10. **NBA 1st-Team:** a cluster of players above average in every statistical variable.

11. **NBA 2nd-Team:** a cluster of players close to average or a little above it in every statistical variable.

12. **Role Player:** a cluster of players slightly less skilled than the NBA 2nd-Team ones, who usually play a few minutes.

13. **One-of-a-Kind:** players that are so good that the algorithm is not able to classify them in any category; they are considered to be just outstanding.

Another study on this topic is proposed by Bianchi et al. (2017), where 5 new positions are defined by means of an integrated approach based on the joint use of a fuzzy clustering technique and a Self-Organizing Map (SOM, Kohonen, 1982, 1990), a machine learning tool consisting of an unsupervised neural network that applies a competitive learning rule (as opposed to error-correction learning, such as for example backpropagation) based on a neighborhood function aimed at preserving the topological properties of the input space. The SOM is fed with 7 selected statistical variables of a set of NBA players, which constitute the input layer of the neural network: PTS (per game), REB ($OREB + DREB$), AST, TOV, STL, BLK, PF. The output layer defines the map space, a two-dimensional region where neurons are arranged in a regular hexagonal or rectangular grid. Each neuron is associated with a weight vector, which defines its position in the input space (the multidimensional space defined by the 7 selected variables). Training consists in moving weight vectors toward the input data, so that the output layer's shape is automatically adapted to the topology of the input space. Thus, the SOM can be considered a mapping from a multidimensional input space to a lower-dimensional map space and is therefore a dimensionality reduction tool. Once trained, an instance from the input space is classified by finding the neuron with the closest weight vector. In Bianchi et al. (2017), the number of neurons of the output layer is set to 900. The output layer is then represented on a plane by Multidimensional Scaling (MDS) and the groups are defined by means of a fuzzy clusterization with a polynomial fuzzifier function applied to the output layer of the SOM. A basketball technical analysis of the cluster profiles, together with the consideration of the membership coefficients of the players to the different clusters, finally allows to classify all the players into the following 5 groups, partly overlapping and confirming the 13 formerly defined by Alagappan (2012):

1. **All-Around All Star:** a cluster of outstanding players who are above the average in most of the game statistics. They are usually elite scorers, but combine the scoring ability with great passing skills; alternatively, they may have big rebounding numbers or great defensive skills; in some cases, all these characteristics are present at the same time.

2. **Scoring Backcourt:** a cluster of players who are characterized by remarkable offensive skills, but usually below the average as for rebounds and blocks. They tend to perform better when playing away from the basket.

3. **Scoring Rebounder:** a cluster of big players who are above the average in scored points (both in low-post and facing the basket) and rebounds.

4. **Paint Protector:** a cluster of players who are not great scorers, but very good at rebounding and blocking shots, thanks to their size and excellent defensive skills.

5. **Role Player:** a cluster of players who are very good but not excellent in only one variable; they are considered as specialists

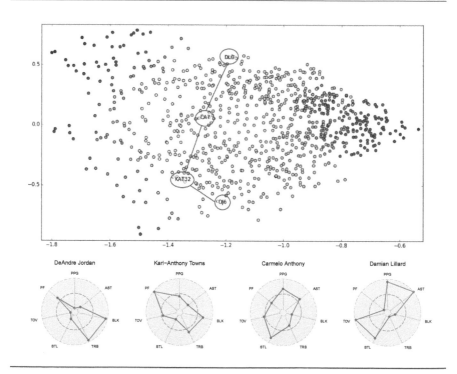

Figure 4.14 Top: positions of four players on the SOM output layer, moving from Paint Protector to Scoring Backcourt region. Bottom: radial plots of the four players. Source: Bianchi et al. (2017).

in one particular aspect of the game. and can be crucial in a team, although their importance is usually not reflected in their statistics, which are usually lower than other players.

The top panel of Figure 4.14 shows the MDS two-dimensional representation of the output layer's neurons, with four players projected from the input space (DeAndre Jordan, Karl-Anthony Towns, Carmelo Anthony, and Damian Lillard), progressively moving from the region where the Paint Protector cluster has been detected, to the Scoring Backcourt cluster region. The radial plots in the bottom panel of Figure 4.14 clearly show how the profiles change moving on the output layer's surface.

The two mentioned studies addressing the issue of defining new roles in basketball make use of complex clustering algorithms, definitely more sophisticated than the k-means and the hierarchical algorithms exploited for the examples in Sections 4.2.1, 4.2.2 and 4.3.1. The reason is that complex algorithms such as the TDA and SOM are able to more effectively explore and preserve the topology of the input space. This allows to detect the possible presence of small groups that differ from others for just one or a few variables, and works well also when the shape of clusters is not hyper-spherical.

GLOSSARY

Cluster Analysis: A classification technique aiming at dividing individual cases (subjects) into groups (clusters) such that the subjects in a cluster are very similar (according to a given criterion) to one another and very different from the subjects in other clusters.

Dendrogram or Cluster Tree: A tree diagram used to illustrate the hierarchical arrangement of the clusters produced by an agglomerative (divisive) algorithm; it shows the sequence of cluster fusion (splitting) and the distance at which each fusion (split) took place.

Explained Variance: A measure for the clusterization quality, understood as the extent to which the individual cases or subjects within the clusters are similar to each other and different from those belonging to other clusters. It is computed as the ratio of the Between over the Total Deviance ($BD/TD = 1 - WD/TD$, Pearson's correlation ratio η^2), usually multiplied by 100 in order to express it as a percentage, where 100% would mean a zero variance within clusters (in each cluster, the cases are all exactly equal to each other). In general, the value has to be interpreted as the percentage of the total variance that can be explained by the clusterization.

Hierarchical Clustering Algorithms: A class of clustering algorithms that seek to build a hierarchy of clusters. They work in a greedy agglomerative or divisive manner by grouping or splitting the clusters one by one, usually based on a distance metric. The agglomerative approach is "bottom up": each subject starts in its own cluster and, moving up the hierarchy, pairs of clusters are progressively merged on the basis of the smallest (nearest) pairwise distance. The divisive approach is "top down": all subjects start in one cluster and, moving down the hierarchy, splits are performed recursively on the basis of the highest (farthest) pairwise distance.

Nonhierarchical Clustering Algorithms: A class of clustering algorithms that seeks to classify subjects into a pre-determined number of clusters, using an iterative algorithm that optimizes a given criterion or loss function. Starting from an initial arbitrary classification, items are moved from one cluster to another, until no further improvement can be made to the criterion value. In general, the achieved solution is not guaranteed to be globally optimal and different initial classifications may lead to different solutions. For this reason, either rules to select how the initial classification is formed or randomization procedures have to be applied.

Self-Organizing Map: An unsupervised neural network that applies a competitive learning rule (as opposed to error-correction learning, such as for example backpropagation) based on a neighborhood function aimed at preserving the topological properties of the input space.

Topological Data Analysis: A class of topological and geometric tools aimed to describe, by means of mathematical, statistical and algorithmic methods, relevant and possibly complex relationships in data structures represented on multidimensional spaces.

CHAPTER 5

Modeling Relationships in Data

S TATISTICAL MODELING is a way to approximate the mechanisms or the rules that govern the functioning of phenomena. It is founded on the assumption that, for any analyzed phenomenon, a given outcome Y is determined by the joint action of a (potentially big) number h of covariates $\mathcal{X} = X_1, \ldots, X_h$ through an unknown function ϕ and a random innovation ε:

$$Y = \phi(\mathcal{X}) + \varepsilon. \tag{5.1}$$

Equation (5.1) represents the so-called "data generating process" (dgp), as the data we observe about the analyzed phenomenon are assumed to come from this mechanism. The hidden set of rules formalized by ϕ and dgp is often referred to by using the metaphor of the "black box", to indicate its complexity and inscrutability. In this context, Y is usually called the dependent variable and the covariates $X_1, \ldots, X_h$ are explanatory or independent variables or predictors. Statistical modeling aims to set up tools able to mimic the functioning of ϕ by means of the information about the dgp contained in a sample of data observed both for the outcome Y and a set of p covariates $X = X_1, \ldots, X_p$ as close as possible to $\mathcal{X}$. In general $\mathcal{X}$ and X do not perfectly overlap because it can occur that (i) not all of the h covariates are known, and/or (ii) for some covariates data are not available, and/or (iii) there are covariates that we consider relevant but they are not. Preliminary procedures of variable selection are usually performed, especially in presence of a big number of candidate predictors, in order to obtain a set X as representative as possible of $\mathcal{X}$.

Fundamentally, there are two goals in developing tools able to approximate the *dgp*: prediction (to be able to predict what the outcome is likely to be for given unknown/future values of the covariates) and information (to extract knowledge about how the outcome and the covariates are associated).

As for prediction, the accuracy of the model is often measured by the coefficient of determination R^2, representing the proportion of the variance in the dependent variable that is predictable by the independent variables. Let $y_1, y_2, \ldots, y_N$ be a sample of N observations of the outcome Y with average $\mu_y = \sum_{i=1}^{N} y_i$, and $\hat{y}_1, \hat{y}_2, \ldots, \hat{y}_N$ the corresponding predictions supplied by a model based on a set of explanatory variables $\mathcal{X}$. We have

$$R^2 = 1 - \frac{SSR}{SST}$$

where

$$SSR = \sum_{i=1}^{N} (y_i - \hat{y}_i)^2$$

is the residual sum of squares, and

$$SST = \sum_{i=1}^{N} (y_i - \mu_y)^2$$

is the total sum of squares. Note that SST corresponds to the Total Deviance TD of Y. The lower SSR, called Residuals Deviance, the lower the differences between real and predicted values, so higher values of R^2 denote better fit of the model/algorithm to the observed data. In some models the coefficient of determination ranges from 0 to 1, and the goodness of fit can be easily interpreted as the percentage of the Total Deviance explained by the model. In other cases R^2 can yield negative values. When it is the case, the simple arithmetic average μ_y is a better predictor than the model. It these cases, interpretation should be very cautious. On the other hand, $R^2 = 1$ always means a perfect fit of predictions to observed data, $\hat{y}_i = y_i \ \forall i$ (which is not necessarily desirable, as it may well be the effect of the so-called overfitting).

Since the *dgp* is unknown and unknowable, according to the assumptions made by researcher about ϕ and ε, several different approaches to statistical modeling can be adopted. An important distinction was given by Leo Breiman (Breiman, 2001b), who claimed that the two main approaches to statistical modeling are the Data Modeling Culture and the Algorithmic Modeling Culture, roughly corresponding to the traditional and the modern concept of modeling.

According to the Data Modeling Culture, ϕ can be represented in an idealized form by a function f indexed by a set of parameters $\boldsymbol{\theta}$ belonging to a parameter space Θ, so that the expected value of Y is given by

$$E(Y) = f(X_1, \ldots, X_p; \boldsymbol{\theta}). \tag{5.2}$$

In this Culture, statistical modeling is a simplified, mathematically formalized way to approximate the elusive black box and the statistical model is the function f that embodies all the assumptions concerning the unknown ϕ governing the functioning of the analyzed phenomenon (see McCullagh, 2002 for a more formal definition). Whilst the mathematical form of f is in general assumed by the researcher, statistical methods are concerned with the estimation of the values of the parameters $\boldsymbol{\theta}$, based on the available sample of data.

The Algorithmic Modeling Culture builds on the fundamental criticism of the mathematical formalization f that, arbitrarily assumed by the researcher, would require more effort to the sophistication of its math, than to the improvement of the model's raw ability in explaining or predicting nature. According to Breiman, the conclusions we draw from this mathematical formalization are geared more towards the model's rather than the nature's mechanism. The Algorithmic Modeling Culture takes a different approach and shifts focus from mathematical models to the properties of algorithms. Instead of developing some elegantly designed model, algorithms try to recreate the black box mechanism by automatically and iteratively finding the way to adapt their output to data. The recent explosion in computing power, the well-known abundance of data and big data, and the creative effort of a big community of statisticians devoted to this new and fascinating branch of research have made the Algorithmic Modeling Culture a powerful approach to statistical modeling. Examples of algorithms developed in this field are trees, ensemble learning algorithms and neural networks, which have been already mentioned in this book, when talking of machine learning tools. A criticism moved against the Algorithmic Modeling Culture is that the mechanisms developed by algorithms to fit data are as inscrutable as the *dgp* itself. In other words, the Algorithmic Modeling Culture favors prediction over information. To make a choice between the Data Modeling and the Algorithmic Modeling Culture means to choose between interpretabiliy and accuracy, a tangible expression of Occam's dilemma between the *lex parsimoniae* (Occam's razor, claiming that simplicity should be preferred in the scientific method), and the evidence that complex models are the most accurate predictors.

A third approach, lying in the middle between the Data Modeling and the Algorithmic Modeling Culture, is the use of nonparametric models, techniques that do not require to specify a model structure *a priori*, but instead determine it from data. The term nonparametric is not meant to imply that parameters are totally absent but that the parameter space is flexible and not defined in advance, in both nature and dimension, and automatically adapts itself to the complexity of the data. Typically, there can be assumptions about the probability distribution of variables and the types of connections among them.

"All models are wrong, but some are useful."

George E. P. Box
One of the great statistical minds of the 20th century

Besides all the statistical and philosophical issues involved in the debate about the best approach to statistical modeling, it should be remembered that it is fundamentally impossible to determine precisely how the conversion from the covariates to the outcome occurs and there will never be an ultimate answer to it. Whether one follows the Data Modeling or the Algorithmic Modeling Culture, or whatever else, there is an inherent limit to the human capacity of explaining nature, due to both high-dimensionality (and then, the impossibility to have exact overlap between $\mathcal{X}$ and X) and the fact that mechanisms are exceedingly complex. We will never reach the exact approximation and knowledge of the *dgp*, we must settle for continuously improving models able to give better and better approximations, which, of course, can be extremely useful.

In the literature, a significant share of the studies dealing with the analysis of basketball data is concerned with the use of statistical modeling techniques. In fact, statistical modeling can be exploited for nearly every analytic purpose listed in Chapter 1, Section 1.2. For the moment, the main part of the proposed models comes from the Data Modeling Culture, but the use of nonparametric techniques and machine learning algorithms is rapidly gaining ground.

In this chapter we will address linear models, the most traditional statistical models in the Data Modeling Culture, and two nonparametric regression techniques. Although the main part of modeling methods is designed to explain multivariate relationships, we will focus on the case of only one explanatory variable and will exploit the R functions (in the package `BasketballAnalyzeR`) `simplereg` performing three types of simple regression analysis, `expectedpts` and `scoringprob`, designed to deal with two specific analytics needs with a nonparametric approach. Some machine learning tools from the Algorithmic Modeling Culture have already been discussed previously, for example, in the FOCUS studies in Chapters 3 and 4.

5.1 LINEAR MODELS

In the Data Modeling Culture, the most traditional functional form for the model is a linear combination of the predictors, implying that a constant change in one explanatory variable (the others being fixed) leads to a constant change in the expected value $E(Y)$ of the outcome:

$$E(Y) = \beta_0 + \beta_1 X_1 + \ldots + \beta_p X_p, \tag{5.3}$$

where $\beta_0, \beta_1, \ldots, \beta_p$ are the model parameters. The model (5.3), called multiple linear regression, can be used when Y is a numerical variable for which we can assume a Gaussian distribution (this assumption is usual but necessary only for inference purposes).

In the linear regression (5.3), when parameters are estimated via ordinary least squares, it can be easily proven that $SSE \leq SST$, so that the coefficient of determination R^2 ranges from 0 to 1, with higher values meaning better fit, and can be interpreted as the percentage of the Total Deviance explained by the models. In the case of only one explanatory variable (simple linear regression), the coefficient of determination is equal to the squared Pearson's linear correlation coefficient, $R^2 = \rho_{XY}^2$. When linear regression is performed without including an intercept, R^2 can yield negative values.

Generalized linear models (GLMs, McCullagh and Nelder, 1989; Fahrmeir and Tutz, 2013) are a flexible generalization of the multiple linear regression model, allowing for a Y variable with a probability distribution other than Gaussian, provided it belongs to the exponential family. In GLMs the linear combination of predictors is related to the expected value of the outcome via the so-called link function g:

$$E(Y) = g^{-1}(\beta_0 + \beta_1 X_1 + \ldots + \beta_p X_p) \tag{5.4}$$

and it allows to deal with several interesting cases, such as the prediction of the probability of a yes/no variable, the number of events occurred for a counting process, phenomenons where an exponential-response mechanism is more appropriate than a linear-response one, etc. Although in the following we will focus on the case of a single predictor variable, we want to stress here the importance of multivariate methods, as in practice variables are always associated to a whole a set of covariates and the use of only one of them may be simplistic in certain cases. However, the multiple linear regression model is a straightforward extension of the simple one that we will describe in the next section. It and its generalizations with GLMs can be very easily applied with R using the `glm` function of the `stats` package.

5.1.1 Simple linear regression model

The model (5.3) with only one predictor variable ($p = 1$) is called simple linear regression and consists of fitting data in the two-dimensional plane by means of a (non-vertical) straight line able to predict, as accurately as possible, the dependent variable values as a function of the independent variable:

$$E(Y) = \beta_0 + \beta_1 X. \tag{5.5}$$

The accuracy is usually measured by the squared residuals (the vertical distances between the points of the data set and the fitted line), and the so-called ordinary least squares method aims to find the line (*i.e.*, to estimate the parameters β_0 and β_1) that makes the sum of these squared deviations as small as possible. When this estimation method is used, the estimated slope $\hat{\beta}_1$ of the fitted line is equal to the correlation between Y and X corrected by the ratio of their standard deviations, and the estimated intercept $\hat{\beta}_0$ is such that the line passes through the point with coordinates given by the averages of the two variables computed on the data points, (μ_x, μ_y):

$$\hat{\beta}_1 = \rho_{XY} \times \frac{\sigma_Y}{\sigma_X} \qquad \hat{\beta}_0 = \mu_y - \hat{\beta}_1 \mu_x. \tag{5.6}$$

For model (5.5) we have $R^2 = \rho_{XY}^2$.

The function `simplereg` allows to fit data with model (5.5) thanks to the parameter `type`, which allows to specify different models (`type = "lin"` for simple linear regression). Since this model makes sense when the variables Y and X have revealed a high linear correlation coefficient ρ_{XY}, we start from the analysis described in Chapter 3, Section 3.2, where we inspected the pairwise linear correlation of the variables scored point, 3- and 2-point shots made, total rebounds (offensive and defensive), assists, turnovers, steals and blocks (per minute played), focusing attention on those players who have played at least 500 minutes in the championship. As shown in Figure 3.2, the variables denoting the number of assists and turnovers per minute exhibited the highest linear correlation coefficient (0.69). So, we fit the model (5.5) with Y and X representing the turnovers and the assists, respectively. It is worth noting that the high correlation between the two variables does not imply any assumption of causal relationship, because correlation is a symmetric association and is not able to reveal which variable (if any) causes the other. Therefore, here, the choice of which variable should be treated as the dependent or the independent one is arbitrary. We perform the analysis through the following code:

```
> Pbox.sel <- subset(Pbox, MIN>=500)
> attach(Pbox.sel)
> X <- AST/MIN
> Y <- TOV/MIN
> Pl <- Player
> detach(Pbox.sel)
> out <- simplereg(x=X, y=Y, type="lin")
> xtitle <- "AST per minute"
> ytitle <- "TOV per minute"
> plot(out, xtitle=xtitle, ytitle=ytitle)
```

The R object `out` of class `simplereg` contains information about the estimated model (parameter estimates, coefficient of determination, residuals, etc.). A plot method is available for this class that produces the graph of Figure 5.1 (top), displaying the scatter plot of the data points, the fitted line, its equation and the value of the coefficient of determination. The value $\beta_1 = 0.26$ indicates that when the number of assists per minute increases by 1, the number of turnovers per minute tends to increase, on average, by 0.26. The values $R^2 = 47.25\%$ denotes a quite good fitting, with the model explaining almost one half the variance of Y. Strictly speaking, the simple linear regression model is based on the assumption that the probability distribution of Y is Gaussian or, at least, that it is continuous with an unbounded domain. In the analyzed case, the outcome variable (turnovers per minute) is countable and expected to be always positive, so that, actually, we ought to avoid the application of a simple linear regression model. However, the flexible criterion

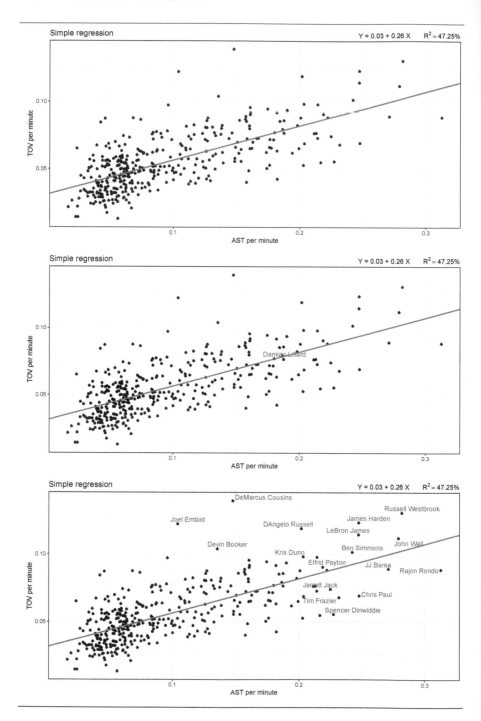

Figure 5.1 Simple linear regression model of assists and turnovers (per minute).

usually adopted in these cases is to fit the model and reject it only when the assumption violation clearly leads to odd results (as happens in the analysis that will be described in Section 5.2.1), which is not the case in the presented example. An alternative in this case would be to resort to some more appropriate generalized linear model.

If needed, we can highlight the position of a specific player (or more than one) we are interested in (Figure 5.1, middle):

```
> selp <- which(Pl=="Damian Lillard")
> plot(out, labels=Pl, subset=selp, xtitle=xtitle,
        ytitle=ytitle)
```

The position of Damian Lillard is just on the line, meaning that he performs, on average, exactly how predicted by the model. We can also require, instead of the position of one or more specific players, the indication of the most outstanding players, specifying the upper and lower quantiles of Y and X we are interested in:

```
> plot(out, labels=Pl, subset="quant",
        Lx=0, Ux=0.97, Ly=0, Uy=0.97,
        xtitle=xtitle, ytitle=ytitle)
```

as shown in Figure 5.1 (bottom), where the players with Y or X higher than the 97th quantile are highlighted. Players lying below the line, *e.g.*, Chris Paul, tend to perform better than what is expected based on the model (in the sense that, for a given number of assists, they have on average a lower number of turnovers with respect to what is predicted by the model); on the other hand, players lying above the line, *e.g.*, DeMarcus Cousins, tend to perform worse than the average from this point of view.

5.2 NONPARAMETRIC REGRESSION

As already pointed out, in nonparametric regression, the relationship between the outcome and the predictor variables is not assumed to take a predetermined form, but is constructed according to information derived from the data. Although the parameter space is not fixed *a priori* and both its nature and size are worked out simultaneously with the model fitting, there may be parametric assumptions about the distribution of model residuals.

Usually nonparametric regression requires a larger sample size than regression based on parametric models, because the data must provide information enough to obtain, alongside the model estimates, the model structure itself. In addition, carefulness is required in interpreting the coefficient of determination R^2, as in some cases with these models it can yield negative values.

Nonparametric regression is typically performed by means of smoothing techniques, such as kernel smoothing, k-nearest neighbor estimates and spline smoothing (Härdle, 1990; Hollander and Wolfe, 1999).

We will focus on kernel smoothing, whose basic idea is to compute the prediction of the outcome as a real-valued function of neighboring observed data, with closer points having a higher impact on the estimated value. Let $y_1, y_2, \ldots, y_n$ be a sample of n observations of the outcome Y and $\boldsymbol{x}_1, \boldsymbol{x}_2, \ldots, \boldsymbol{x}_n$ the corresponding observed values for the explanatory variables $\boldsymbol{X}$, the simplest form of kernel estimate is the Nadaraya-Watson weighted average

$$\hat{y}_0 = \frac{\sum_{i=1}^n K_\lambda(\boldsymbol{x}_0, \boldsymbol{x}_i) y_i}{\sum_{i=1}^n K_\lambda(\boldsymbol{x}_0, \boldsymbol{x}_i)} \tag{5.7}$$

where $\hat{y}_0$ is the outcome value predicted at $\boldsymbol{X} = \boldsymbol{x}_0$ and $K_\lambda(\boldsymbol{x}_0, \boldsymbol{x}_i)$ is the so-called kernel, a window function that defines the weights to be given to the points in the neighborhood of $\boldsymbol{x}_0$ and depends on a smoothing parameter λ, controlling the width of the neighborhood, that in this context is usually called bandwidth. We already came across the concept of kernel in Chapter 3, Section 3.5, with reference to density estimation procedures. Different specifications of the kernel lead to different smoothers, the most popular one being the Gaussian kernel smoother.

A more general approach than the simple Nadaraya-Watson weighted average is local regression, where the prediction is obtained by means of some parameterized function, such as a low-order polynomial, with parameters estimated by minimizing a least squares loss function where closer points are given higher weights (see Friedman et al., 2009 for details). When the fitted polynomials have degree 0, the method leads back to (5.7).

In the following, we will describe some examples where nonparametric regression is carried out on basketball data using a polynomial local regression technique and Nadaraya-Watson Gaussian kernel smoothing.

5.2.1   Polynomial local regression

The term "polynomial local regression" is usually referred to a nonparametric technique called LOESS, introduced by Cleveland (1979) and

Cleveland and Devlin (1988). LOESS fits simple parametric models to localized subsets of the data so as to obtain a smooth curve. In detail, at each point of the dataset, a low-degree polynomial is fitted to a subset of the data in the neighborhood of the point itself. The polynomial parameters are determined by a weighted least squares method, giving more weight to points closer to the point whose outcome is being estimated. The outcome prediction is then obtained by evaluating the local polynomial using the explanatory variables values for that data point. Although there are some typical default choices, the researcher can tune the degree of the polynomials, the weights and the smoothing parameter λ.

With respect to parametric models, LOESS requires a significantly increased computational burden, so that it would have been practically unfeasible in the era when least squares parametric regression was being developed. The improved computational power has made it possible to design LOESS, as well as many other computationally intensive new methods, to achieve goals not easily gotten by traditional approaches.

The level plots of Figure 3.5, produced by the function `MDSmap` described in Chapter 3, Section 3.3, are obtained by performing a LOESS bivariate regression with second-degree polynomials, where the explanatory variables are the MDS coordinates. In this section, instead, we focus on the case of only one explanatory variable. The function `simplereg` allows to perform simple regressions with LOESS by specifying `type = "pol"`. When using LOESS, `simplereg` is based on the R function `loess`; therefore, it has all its default settings. The polynomials' degree is set to 2 and the smoothing parameter can optionally be tuned by the researcher using the argument `sp`, denoting the size of the subset used to fit each local polynomial, expressed as a fraction of the total number of data points.

Like in Section 5.1.1, we start from the analysis of the pairwise linear correlation presented in Chapter 3, Section 3.2, where we detected a moderately high negative linear correlation coefficient (-0.52) between rebounds and 3-point shots made (Figure 3.2). We first prepare the data with the following code

```
> Pbox.sel <- subset(Pbox, MIN>=500)
> attach(Pbox.sel)
> X <- (DREB+OREB)/MIN
> Y <- P3M/MIN
> Pl <- Player
> detach(Pbox.sel)
```

Then, we note that the presence of linear correlation would suggest to use a linear regression model, which, however, in this case, does not provide an adequate fit to data, as shown in the top panel of Figure 5.2, obtained by typing the code

```
> out <- simplereg(x=X, y=Y, type="lin")
> xtitle <- "REB per minute"
> ytitle <- "P3M per minute"
> plot(out, xtitle=xtitle, ytitle=ytitle)
```

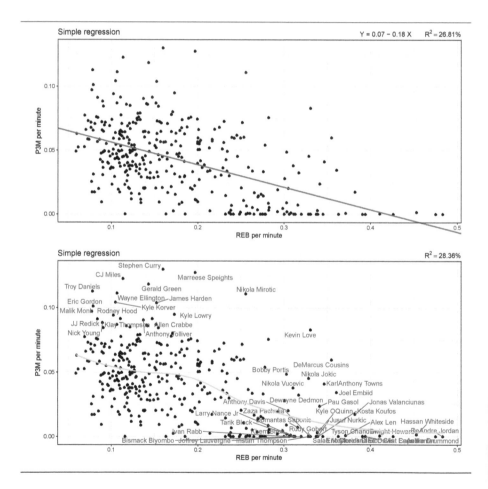

Figure 5.2 Top: Simple linear regression model of rebounds and 3-point shots made (per minute). Bottom: LOESS regression of rebounds and 3-point shots made (per minute).

It shows that the model turns out to predict negative values for 3-point shots made for those players having the highest rebound values. This is a consequence of the above-mentioned (Section 5.1.1) violation of the assumptions, required by the simple linear regression model, on the probability distribution of Y. So, we perform a LOESS regression, and require to specify in the plot the names of the players lying above the 90th and 95th quantile for X and Y, respectively (Figure 5.2, bottom)

```
> out <- simplereg(x=X, y=Y, type="pol")
> plot(out, labels=Pl, subset="quant",
       Lx=0, Ux=0.90, Ly=0, Uy=0.95,
       xtitle=xtitle, ytitle=ytitle)
```

According to the LOESS curve, the expected number of 3-point shots made is approximately flat or very slowly decreasing for players with less than about 0.2 rebounds per minute (9.6 rebounds in 48 minutes), and then tends more rapidly to zero. Players with more than 0.3 rebounds per minute (14.4 rebounds in 48 minutes) are expected to score very few 3-point shots (less than 0.02 per minute, 0.96 in 48 minutes), but the data points are very scattered in that area, so that we can recognize a number of outstanding players. The most noticeable one is Kevin Love who, although achieving 0.33 rebounds per minute (15.84 in 48 minutes), scores more than 0.08 3-point shots per minute (3.84 in 48 minutes). Other players worth mentioning are DeMarcus Cousins, Bobby Portis, Nikola Jokic and Karl-Anthony Towns. On the other hand, in the top part of the scatter, we find outstanding players from the point of view of the 3-points shots made. Among them, the one achieving the highest number of rebounds is Nikola Mirotic, with reference to the 622 minutes, 25 games, he played with the Chicago Bulls (0.26 rebounds and 0.11 3-point shots made per minute, corresponding to 12.48 and 5.28 in 48 minutes, respectively). The coefficient of determination is rather low (28.36%, but still better than the value obtained with the simple linear regression, 26.81%), meaning that predictions are affected by high variability.

5.2.2 Gaussian kernel smoothing

In Gaussian kernel smoothing, the kernel $K_\lambda(\boldsymbol{x}_0, \boldsymbol{x}_i)$ is given by

$$K_\lambda(\boldsymbol{x}_0, \boldsymbol{x}_i) = \frac{1}{\lambda} \exp\left\{ -\frac{||\boldsymbol{x}_i - \boldsymbol{x}_0||^2}{2\lambda} \right\} \tag{5.8}$$

where $|| \cdot ||$ denotes a norm function and the smoothing parameter λ corresponds to the variance of the Gaussian density. In the case of only one explanatory variable, the function `simplereg` can be used to perform the Nadaraya-Watson smoothing (5.7) with the Gaussian kernel (5.8), by specifying `type = "ks"`.

Kernel smoothing is also performed by the function `scatterplot`, introduced in Chapter 2, Section 2.2.4, again used in Chapter 3, Sections 3.2 and 3.5.3 to analyze, respectively, linear correlation and joint densities of pairs of variables within a given set. Let us select those players who have played at least 500 minutes and consider the following variables: scored point, 3- and 2-point shots made, total rebounds (offensive and defensive) and assists (per minute played):

```
> data <- subset(Pbox, MIN>=500)
> attach(data)
> X <- data.frame(PTS, P3M, P2M, REB=OREB+DREB, AST)/MIN
> detach(data)
```

We can plot a correlation matrix with histograms on the diagonal and kernel smoothing estimates of the bivariate relationships in the lower triangle.

```
> scatterplot(X, data.var=1:5,
              lower=list(continuous="smooth_loess"),
              diag=list(continuous="barDiag"))
```

The outcome is shown in Figure 5.3.

The Nadaraya-Watson Gaussian kernel smoothing is also used in the functions `scoringprob` and `expectedpts`, estimating, respectively, the scoring probability and the expected points of a shot as a function of a game variable. In detail, given a game variable X (for example, the distance from the basket, or the time in the quarter when the shot is attempted, etc.), the scoring probability $\pi(x)$ denotes the probability that a shot with $X = x$ scores a basket, while the expected points $E(PTS|x)$ inform on the number of points scored on average thanks to a shot with $X = x$. The fine line between these two concepts can be explained by a numerical example. Let us consider Kyrie Irving, having $P2p = 54.12$ and $P3p = 40.79$. Given a distance x such that the shot is worth 2 points, we have $\pi(x) = 0.5412$ because the scoring probability can be roughly estimated by the scoring percentage at x and $E(PTS|x) = 2 \cdot 0.5412 = 1.0824$, as the expected points are given by the number of points in case of shots made, multiplied by the scoring probability. Given a distance x such that the shot is worth 3 points, we have $\pi(x) = 0.4079$ and $E(PTS|x) = 3 \cdot 0.4079 = 1.2237$. This means that,

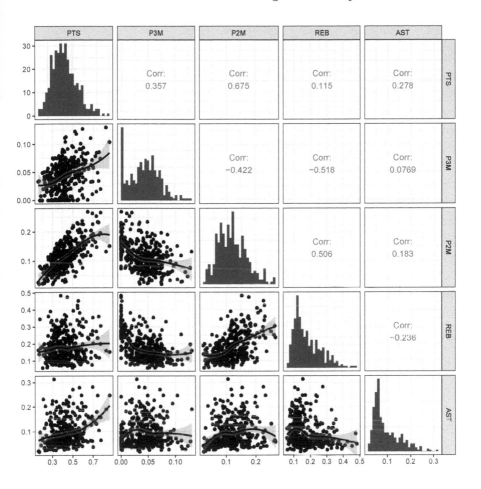

Figure 5.3 Correlation matrix with histograms on the diagonal and kernel smoothing estimates of the bivariate relationships in the lower triangle.

although Kyrie Irving's scoring probability is higher for 2-point than for 3-point shots, the expected points (*i.e.*, the number of points we expect he scores on average when the shot distance is x) are higher when he shoots from behind the 3-point line. With the functions `scoringprob` and `expectedpts`, $\pi(x)$ is estimated with a Gaussian kernel smoothing that, with respect to the simple example we have described, is more accurate and provides point estimates for all the possible values of the variable X.

Both the functions need play-by-play data in input and the smoother bandwidth can be controlled with the parameter `bw`. For the examples, we use the play-by-play dataset `PbP`, obtained from `PbP.BDB` thanks to the function `PbPmanipulation` as described on page 23.

5.2.2.1 Estimation of scoring probability

With `scoringprob` we assess the scoring probability as a function of the period time (`periodTime`, seconds played in a given quarter), the total time (`totalTime`, seconds played in total), the play length (`playlength`, time between the shot and the immediately preceding event), and the shot distance (`shot_distance`, distance in feet from the basket). The variables are the same used in Chapter 3, Section 3.5.1 to analyze the pattern of the shooting density. Now, as then, the idea comes from the words of coach Messina, who remarked in the Foreword that, "*... scoring ten points in the first two quarters of the game is not the same as scoring them in the final minutes...*", thus pointing out the importance of investigating players' performance with respect to some specific situations, and not only on average.

We now estimate the scoring probability of the Golden State Warriors' 3-point shots with respect to play length and period time (note that for the latter we need to tune the bandwidth using the argument `bw`):

```
> PbP.GSW <- subset(PbP, team=="GSW" & result!="")
> p1 <- scoringprob(data=PbP.GSW, shot.type="3P",
                    var="playlength")
> p2 <- scoringprob(data=PbP.GSW, shot.type="3P",
                    var="periodTime", bw=300)
> library(gridExtra)
> grid.arrange(p1, p2, ncol=2)
```

obtaining the graphs of Figure 5.4. The scoring probability tends to decrease when the buzzer sound (both for the shot clock and the game clock) approaches. The estimated patterns are consistent with the evidences found by Zuccolotto et al. (2018), who investigated the effects of shooting in high-pressure game situations, using data from completely different championships (the Italian "Serie A2" and the Olympic tournament Rio 2016). The function also allows to estimate the scoring probabilities of specific players:

```
> pl1 <- c("Kevin Durant","Draymond Green","Klay Thompson")
> p1 <- scoringprob(data=PbP.GSW, shot.type="2P",
                    players=pl1, var="shot_distance",
                    col.team="gray")
> pl2 <- c("Kevin Durant","Draymond Green")
> p2 <- scoringprob(data=PbP.GSW, shot.type="2P",
                    players=pl2, var="totalTime", bw=1500,
                    col.team="gray")
```

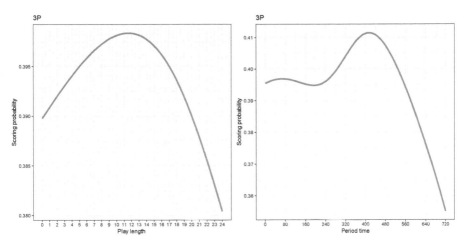

Figure 5.4   3-point shots scoring probability with respect to play length and period time.

```
> library(gridExtra)
> grid.arrange(p1, p2, ncol=2)
```

Observing the graphs of Figure 5.5, we draw the following remarks:

- the scoring probability of 2-point shots decreases as the shot distance increases;

- Kevin Durant performs better than the team average from all the distances;

- Draymond Green's decrease is steeper than the average for distances higher than about 7 feet;

- Klay Thompson performs appreciably worse than the average from short distances, but his scoring probability becomes approximately flat from about 12 feet, and from long distances he achieves the best performance with respect to both the average and the other two teammates;

- the team scoring probability of 2-point shots peaks in the third quarter of the game and then decreases as the end of the game approaches;

- Kevin Durant tends to perform better in the first half of the game;

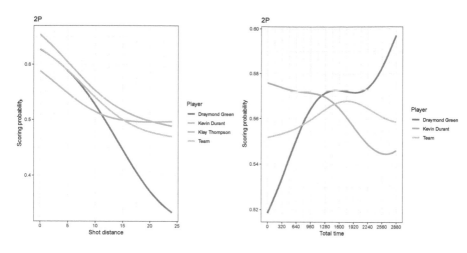

Figure 5.5   2-point shots scoring probability with respect to shot distance and total time.

- Draymond Green tends to perform better in the second half of the game.

5.2.2.2   Estimation of expected points

With `expectedpts`, we can estimate the expected points with respect to the same game variables usable in `scoringprob` and compare, in a unique graph, the performance of single players, also with respect to the team average. This allows to determine, for each player, the circumstances when his shots have the maximum efficiency, taking into account both the points brought by their shots and their scoring probability. When the game variable is the distance of the shot from the basket (default), the analysis informs about the best distance (for each player) to shoot from. For example we may be interested in Stephen Curry and Kevin Durant: with the following code lines, we obtain the plot in the top panel of Figure 5.6.

```
> PbP.GSW <- subset(PbP, team=="GSW")
> pl <- c("Stephen Curry","Kevin Durant")
> mypal <- colorRampPalette(c("red","green"))
> expectedpts(data=PbP.GSW, players=pl,
              col.team="gray", palette=mypal,
              col.hline="gray")
```

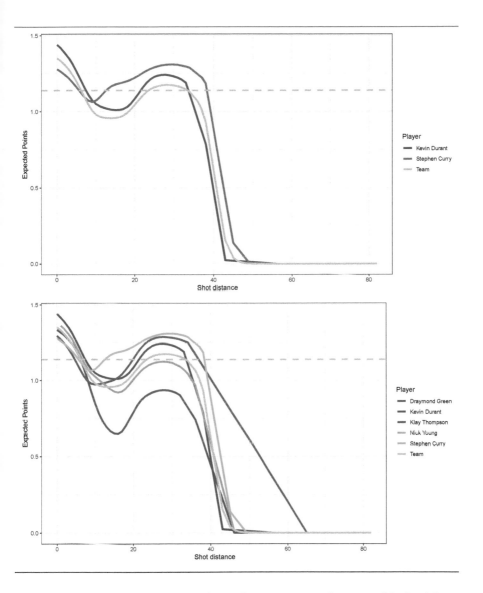

Figure 5.6 Expected points of shots from a given distance (dashed line: team average independently from shot distance). Top: Comparison between Stephen Curry and Kevin Durant. Bottom: Analysis of all the Golden State Warriors players who scored more than 500 points.

By observing the graph, we derive the following remarks:

- for shots attempted from less than about 10 feet, Kevin Durant performs better and Stephen Curry performs worse than the team average;

- for shots attempted from more than about 10 feet, Kevin Durant is still better than the team average, but Stephen Curry is better than Kevin Durant;

- Stephen Curry achieves his best performance in terms of expected points when shooting from a distance higher than 22 feet (*i.e.*, with 3-point shots); he should avoid taking it to the rim; nevertheless, for middle distance 2-point shots (more than about 10 feet) he performs better than the team average;

- Kevin Durant achieves his best performance in terms of expected points when shooting from the restricted area (less that 4 feet); he is nevertheless always better than the team average, except for shots from very far away (more than 35 feet).

This graph also confirms the well-known inefficiency of shots from the middle distance, also mentioned by Ettore Messina in the Foreword of this book. Nevertheless, he also claimed that, "*if all those in attack go for shots from below and shots from the three point line, the defenders will do their best not to concede these shots. As a result, they may allow more shots from the middle. At this point, having some athletes in the team who can shoot well from this distance will become absolutely fundamental in order to win...*". Suppose we wish to know who could such a player be in the Golden State Warriors. We first select those players who have scored at least 500 points in the regular season, then we generate the plot of expected points (Figure 5.6, bottom), where we find that Stephen Curry's efficiency from the middle distance is appreciably higher than the team average (and even almost the same as his efficiency from behind the 3-point line), so he could be the one to rely on in case of application of the strategy suggested by coach Messina.

```
> Pbox.GSW <- subset(Pbox, PTS>=500 &
                     Team=="Golden State Warriors")
> pl <- Pbox.GSW$Player
> mypal <- colorRampPalette(c("red","green"))
> expectedpts(data=PbP.GSW, players=pl,
```

```
            col.team="gray", palette=mypal,
            col.hline="gray")
```

For the same players, we may investigate the expected points of their shots with respect to the time in quarter when the shot is attempted. In this case, the expected points are affected by the specific combination of 2-point and 3-point shots (that could vary depending on the time in the quarter), as well as by the scoring probability (separately for 2-point and 3-point shots, which could also vary in time). Note that in this case we have to tune the bandwidth by means of the argument bw

```
> expectedpts(data=PbP.GSW, bw=300, players=pl,
            col.team="gray", palette=mypal,
            col.hline="gray", var="periodTime",
            xlab="Period time")
```

The result is displayed in Figure 5.7, which suggests several interesting remarks:

- Stephen Curry exhibits the highest efficiency, with just a slight decrease in the first half, and definitely the most reliable in the last minutes;

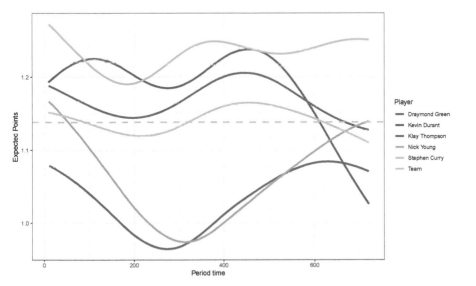

Figure 5.7 Expected points of shots with respect to the period time (analysis of all the Golden State Warriors players who scored more than 500 points).

- Klay Thompson and Kevin Durant perform better than the average, but seem to suffer a downswing in the last minutes;

- Draymond Green and Nick Young appreciably improve their expected points in the second half of the quarter.

Lastly, we would like to address another important point raised by coach Messina in the Foreword: "*Too many players that average ten points per game, for the season, score most of those points against average and poor teams, whereas they are less successful against the top teams*". Along with other methods presented throughout the book, the estimation of expected points is a good way to investigate the different performance of players against different opponents. In the following, we will consider the games played by the Golden State Warriors against teams who qualified for the Playoffs (object top), separately from those played against the others (object bot)

```
> top <- subset(Tadd, Playoff=="Y" & team!="GSW")$team
> bot <- subset(Tadd, Playoff=="N")$team
```

In order to streamline the operations, we define a function aimed to select the rows of PbP corresponding to games played against an opponent belonging to a given set. In addition, we modify the players' names by adding a suffix denoting to which set the opponent belongs.

```
> bot_top <- function(X, k) {
          dts <- subset(subset(X, oppTeam %in% get(k)),
                        team=="GSW")
          dts$player <- paste(dts$player, k)
          return(dts)
          }
```

Thanks to the function bot_top we extract and label players' names with respect to the two sets of teams in top and bot, and we combine them by rows in order to generate the input data frame for the function expectedpts, computed in this case with respect to the total time

```
> PbP.GSW <- rbind(bot_top(PbP, "top"),
                   bot_top(PbP, "bot"))
> pl <- c("Stephen Curry top","Stephen Curry bot",
          "Kevin Durant top", "Kevin Durant bot")
> mypal <- colorRampPalette(c("red","green"))
> expectedpts(data=PbP.GSW, bw=1200, players=pl,
```

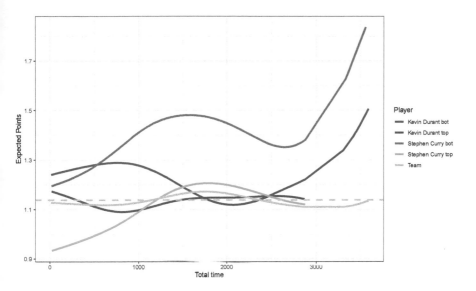

Figure 5.8 Expected points of shots with respect to total time (against top- and bottom-ranked opponents).

```
col.team="gray",   palette=mypal,
col.hline="gray", var="totalTime",
xlab="Total time")
```

The resulting graph (Figure 5.8) shows that both Stephen Curry and Kevin Durant have an appreciably different performance against top- and bottom-ranked opponents[1]. In detail, the number of expected points against bottom-ranked opponents is outstandingly high, especially for Stephen Curry. The story is completely different against top-ranked opponents, which force them to a more ordinary performance.

5.3   FOCUS: SURFACE AREA DYNAMICS AND THEIR EFFECTS ON THE TEAM PERFORMANCE

Statistical modeling is a hugely wide world, composed of hundreds of different models, suited to fit specific data or situations. The models

---

[1]Note that the different extension of the curves derives from the fact that all the games against top-ranked opponents lasted 48 minutes (2880 seconds, the regular time), while against bottom-ranked opponents it happened twice to need overtime (in both cases against the Los Angeles Lakers). The sharp increase of the curves in that period suggests a markedly improved effort of the two players in that tricky situation.

examined in the previous sections are but a few examples of the most traditional and simple ones. A complete listing of all the models that may be used with basketball data is virtually impossible.

This section is devoted to describe just one approach, proposed by Metulini et al. (2018), where some very specific statistical models are used in order to characterize the dynamics of surface areas during a game and detect their influence on team performance. The distinctive features of the case study we are going to present lie in (*a*) the use of sensor data recorded with GPS technologies, and (*b*) the transmigration into this context of statistical models typically employed in other application fields.

Regarding point (*a*), data refer to players' coordinates collected during three matches played in February 2017 by Italian professional basketball teams, at the Italian Basketball Cup Final Eight. The players' positions (in pixels of $1m^2$) in the x- and the y-axis of the court, as well as in the z-axis (*i.e.*, how high the player jumps) were recorded thanks to microchips worn by each player, with an average frequency of about 80 Hz. The system recorded a series of 4,733,124 observations for the first match (hereafter, CS1), 4,072,227 for the second match (CS2) and 4,906,254 for the third one (CS3), each one referring to one among positioning, velocity or acceleration in one among x-, y- or z-axis, for a specific player in a specific time instant. These observations have been transformed into a time series of convex hulls areas regularly spaced in time, with a constant frequency of 10 Hz. A convex hull of a set of points in the Euclidean plane is the smallest convex set that contains all the points. Roughly speaking, if the points were spikes on a wall, the convex hull could be visualized as the shape enclosed by a rubber band stretched around them. As suggested by Passos et al. (2016), the convex hull areas are used as a measure of the effective playing space, called surface area in the sports analytics literature.

When it comes to point (*b*), the basic assumption in the paper is that the surface areas' stochastic process is affected by recurrent structural changes involving its mean. This idea comes from sports technical considerations: it is an established fact that the space among players tends to switch from narrow to large when moving from defense to offense phases (Figure 5.9). This is also supported by statistical evidence in the three examined case studies, where the median values of the surface areas are found to lie in the range of m^2 22-25 and 44-52, for defense and offense, respectively.

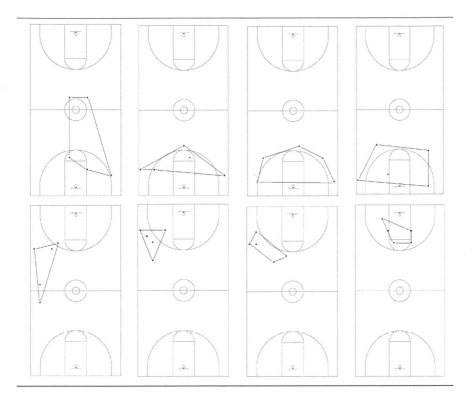

Figure 5.9 Examples of convex hulls of offense (top) and defense (bottom) phases. Source: Metulini et al. (2017a).

Nevertheless, we cannot assume a strict matching between the surface area and the game phase. Quite the opposite, we are just interested in the moments when the surface area is different from what we would expect on the basis of the game phase, because the most intriguing game situations are hidden right there. In order to investigate this issue, the authors needed a tool that is able to detect the switching of surface areas from large to narrow (and vice versa) without referring to the game phase. To do that, statistical models were borrowed from the fairly different context of econometrics, where time series with structural changes are often analyzed with regime-switching models. In that context, regimes are defined as states involving different parameters for the stochastic process under study, and they are often found to correspond to specific economic situations (*e.g.*, expansion or recession). Translating the idea into basketball, the authors assumed that surface area dynamics are characterized by different regimes involving different mean levels of the process. So, the time series of convex hull areas is fit with a Markov Switching Model (MSM; see Hamilton, 2010), able to detect if a regime-switching dynamic is present, and estimate the parameters of the process in the different regimes and, for each observation time, the probability of being in one regime or the other. Finally, the authors investigated the relationship between the regime probabilities and the scored points by means of Vector Auto Regressive (VAR; see Zha, 2010) models.

In all the tree case studies, the MSM detected a significant presence of two regimes, characterized by average surface areas in the range of about m^2 21.7-24.7 and 55.7-63.3 for the regime with narrow and large convex hulls, respectively. In addition, in all the cases, the regimes switch very quickly from one to another (the average duration is 7-8 seconds), which suggests that the two regimes do not perfectly match the offensive and defensive game phases. The association between regimes and game phases is quite strong, as suggested by the Cramer's V index (see Chapter 3, Section 3.1.1) ranging from 43.95% to 56.22% in the three case studies, but not huge. This means that large (narrow) surface areas occur even during defense (offense) and this could have an effect on the team performance.

Before turning attention to the measurement of this effect, the paper investigates whether the presence of a given player or lineup on the court affects surface areas, separately for offense and defense phases. A set of preliminary remarks can be drawn by observing the relative frequencies of the two regimes when each player is on the court or on the bench. Just by way of an example, in case study CS1 the relative frequency of the regime with narrow surface area during offense is 0.303 when player $p1$ is on the bench. When the same player is on the court, the same relative frequency decreases to 0.207, meaning that, on average, the team tends to play more spread in offense when player $p1$ is on the court. Similar considerations apply to the other players and the other case studies. The authors introduce two functions, $\Phi_D^{(L)}(t)$ and $\Phi_O^{(L)}(t)$, obtained by means of a kernel smoothing procedure applied to the probabilities of being in the regime with large surface areas, separately for defense (subscript D) and offense (subscript O) game phases. Plotting these functions with superimposed gray areas corresponding to the presence on the court of a specific lineup or player allows to inspect the regime's patterns during offense and defense crossed with the selected player or lineup.

In Figure 5.10 we may notice that the offensive play until the first half of the second quarter (around minute 15) seems to be apprecia-

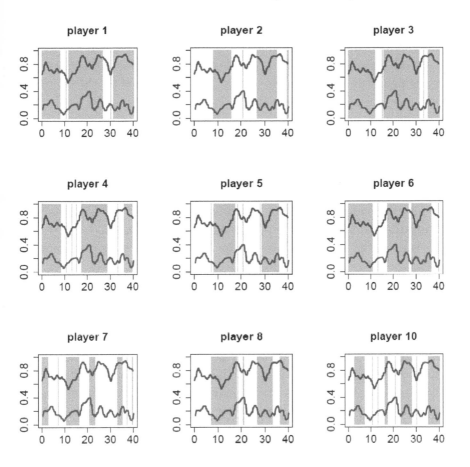

Figure 5.10 Case study CS1: pattern of the functions $\Phi_D^{(L)}(t)$ (blue) and $\Phi_O^{(L)}(t)$ (red) in time (in minutes, x-axis). For each player, gray areas denote the moments when he was on the court. Source: Metulini et al. (2018).

bly different from what follows, as the kernel smoothed probabilities $\Phi_O^{(L)}(t)$ of the regime with large surface areas are considerably lower in the first 15 minutes than in the rest of the match. In addition, something happened in the offensive play around minute 30, as the function $\Phi_O^{(L)}(t)$ clearly falls. Similar remarks can be done with reference to defensive phases: a peak is evident in the function $\Phi_D^{(L)}(t)$ in the middle of the match, around minute 20. These observed fluctuations can be related to the presence of a specific player on the court by looking at the superimposed gray areas. These charts allow a deeper knowledge of the surface area dynamics during the match and with reference to the lineups and players. In addition, further relevant game variables (e.g., the implementation of playbooks, the coach's judgments on the technical performance during the match, etc.) could be added in order to inspect their association with the surface area dynamics.

At this step, we need to assess whether the observed fluctuations of the regime probabilities have a positive or negative impact on the overall team performance. As mentioned above, to this aim, the authors propose the use of VAR, a class of statistical models used to capture the linear interdependencies among multiple time series. In VAR models, each variable has an equation explaining its evolution based on its own lagged values, the lagged values of the other model variables, and an error term. Also this class of models is traditionally applied in the field of macroeconomic econometrics where, in turn, it has arrived, advocated by Sims (1980), coming from system identification and control theory. This completely justifies the translating of the VAR models into sports analytics applications. In the examined case studies, bivariate VAR models were fitted to data, with variables given by the regime probabilities and the points scored by the team (in the analysis of offensive phases) or the opponent (in the analysis of defensive phases). While the evidences found in the first part of the paper were basically common to all the three analyzed case studies, the features emerging from this last investigation are match-specific, which appears to be reasonable because of the different tactics decided by coaches or the different ways of playing determined by the interaction of the two specific teams involved in the match. In two case studies (CS1 and CS3) a significant positive influence of a large surface area in offense on the points scored by the team was detected. In CS2 the model revealed a negative correlation between the points scored during one minute on the points scored in the following one, probably due to a high reliance of the team on 3-point shots that are more volatile than other kinds of shots. In CS1 and CS2, emerged a weak negative effect of the points made by the opponent on the probability to be in the regime with narrow surface areas in defense, as if a more efficient game of the opponent (in terms of scored points) had forced the team to keep the defense spread.

The results presented in the examined paper could be used by basketball coaches and experts as they relate with tactics, specifically with the choice of game strategies, players and lineups. The main findings of the presented case study can be summarized as follows:

- there are robust evidences of the presence of structural changes in the average surface area;

- the presence of certain players or lineups is associated to different probabilities to be in the two regimes;

- there exist some match-specific causal relationships between surface areas and team performance, which have to be interpreted with reference to the specific characteristics of each match.

GLOSSARY

Black box: In general, the term black box is used to denote a device or a system which is only evaluated in terms of its inputs and outputs, without any knowledge of its internal working. In Data Science, it is usually referred to the data generating process that is being emulated by a statistical model or the hidden mechanism of some machine learning algorithms, such as neural networks or ensemble learning procedures, that make predictions without returning, as a result, the explicit functional form used to compute them.

Coefficient of determination: In statistical modeling, the coefficient of determination, usually denoted as R^2, is the proportion of the variance in the dependent variable that is explained by a given model based on a set of independent variables and provides a measure of how well that model is able to predict the outcome given the explanatory variable values. Roughly speaking, it may also be considered as a measure of the prediction improvement given by the model with respect to the poorest predictor, the simple arithmetic average of Y. In linear regression, when parameters are estimated via ordinary least squares, it can be easily proven that the coefficient of determination ranges from 0 to 1, with higher values meaning a better fit. In the case of only one explanatory variable (simple linear regression), the coefficient of determination is equal to the squared Pearson's linear correlation coefficient. In other cases, such as, for example, when linear regression is performed without including an intercept, when the model implies fitting nonlinear functions to data, with nonparametric models and with many models from the Algorithmic Modeling Culture, R^2 can yield negative values. When it is the case, the simple arithmetic average $\bar{y}$ is a better predictor than the model.

Data generating process: A data generating process (dgp) is the hidden set of rules, usually complex and inscrutable, governing the way phenomena occur, with a focus on their observable and measurable expressions (data) and, specifically, on the mechanism according to which a specific outcome Y derives from the joint effect of a number of covariates $X_1, X_2 \ldots, X_h$. In statistics the dgp is assumed to involve both a systematic and a random component, according to a nondeterministic approach to the functioning of Nature.

Overfitting: The term overfitting (or overtraining) describes the particular situation of a prediction model whose output corresponds too closely to the data used to build the model itself, and may therefore fail to fit additional observations or predict future values reliably. The essence of overfitting is to have modeled the noise as if it was part of the data generating process. In general, this can occur when the model contains more parameters than can be justified by the data (as an extreme example, think to the case when there are as many parameters as the sample size) and/or it has such a complex structure to be able to follow very closely even little variations in data. The risk of overfitting is particularly high with models from the Algorithmic Modeling Culture, *i.e.*, in the context of machine learning. In order to avoid overfitting, the model selection should be done by measuring goodness of fit on a set of different data with respect to those used for building the model itself. See Sarle (1996), Sollich and Krogh (1996) and Hawkins (2004) for further discussion.

Smoothing: To smooth a data set means defining an approximating function aimed at capturing important patterns in the data, while leaving out noise or other fine-scale effects. In general, smoothing is used to provide analyses that are both flexible and robust, as usually it requires neither strong distributional assumptions for data, nor *a priori* hypotheses about an explicit function form for the relationships between outcome and predictors. Many different algorithms are used in smoothing, the most popular classes being kernel smoothing, k-nearest neighbor estimates and spline smoothing (Härdle, 1990; Hollander and Wolfe, 1999). From the absence of any functional assumption, it follows that the immediate result from smoothing are the smoothed values, *i.e.*, the expected values for Y corresponding to the explanatory variables' values used for smoothing, which can serve to plot the so-called smoothing curve. Even in the case when the smoothing procedure itself implicitly arranges some functional form, it cannot be used later for prediction when the explanatory variables' values are outside the range of the observed values used for smoothing. The smoothing procedure often depends on a tuning parameter, called smoothing parameter or bandwidth in the context of kernel smoothing, used to control the degree of smoothing, in the sense that they allow to obtain a more or less regular smoothing curve.

III

Computational Insights

CHAPTER 6

The R Package BasketballAnalyzeR

Marco Sandri
BODaI-Lab (**B**ig & **O**pen **Da**ta **I**nnovation **L**aboratory),
University of Brescia

6.1 INTRODUCTION

BasketballAnalyzeR is an open-source package for the statistical language R, designed for the analysis and visualization of basketball data. The package takes advantage of the powerful graphical abilities added to R by the ggplot2 package and allows to produce publication-quality graphics with minimal effort (Murrell, 2016; Wickham, 2016).

The aim of BasketballAnalyzeR is twofold:

- *Simplicity.* Building some of the advanced graphs presented in this book can be a difficult task for a novice R user, because it requires a good understanding of the ggplot2 philosophy and syntax and of other packages that extend ggplot2 graphics capabilities (for example, ggnetwork, ggrepel, GGally). The functions implemented in our package have been designed to minimize the effort required to the user in performing the statistical analysis and the graphical visualizations proposed in the book. The only real difficulty that must be faced by the reader is formatting the input dataset(s) according to the structure required by the functions in the package. To this purpose, in the next section, we illustrate with an example how to transform publicly available datasets into a format suitable to be inputted into the function of our package.

- *Flexibility.* For each function of our package, we implemented a set of options that allow partial customization of the analysis according to the user's requirements. Many of these options can be ignored by the standard user who needs only to replicate the analysis presented in the book. Only more advanced users can take advantage of these additional settings. In addition, all the plotting commands available in our package return a `ggplot` graphical object (or a list of `ggplot` objects). An experienced user can subsequently customize several characteristics of these objects: modify legend, plot background and border, major and minor grids, axis text and ticks, and more. Points, lines, text, labels, annotations, and images can also be added to the original plot and two or more graphical objects can be arranged into a grid drawing multiple plots on a single page.

The package must be first installed following the instructions available at

`https://bdsports.unibs.it/basketballanalyzer/`

This webpage, specifically devoted to the package and addressed to the readers of this book, is continuously updated. It has to be considered a reference point by all the users of `BasketballAnalyzeR`, who can find codes, news about the package, possible updates, discussions about data preparation and contact information for any question they may want to ask the developers.

The command `library(help="BasketballAnalyzeR")` shows basic information and a list of functions and datasets available with the package. The `help()` function and `?` help operator provide access to the documentation pages. For example, `help("plot.assistnet")` and `?plot.assistnet` show the help page of the plot command for the `assistnet` objects. Help pages for functions include a section with executable examples illustrating how the functions work. These examples can be executed via the `example()` command; e.g., `example("plot.assistnet")`.

The next sections describe how to prepare data (Section 6.2), customize plots (Section 6.3) and build interactive graphics (Section 6.4).

All the analyses involving data or other information downloadable from websites refer to versions retrieved on 31st March 2019.

6.2   PREPARING DATA

The aim of this section is to show, to readers with only a basic knowledge of the R language, how `BasketballAnalyzeR` can be used for the statistical analysis of their dataset(s). The easiest and safest way for a correct replication on an arbitrary dataset of the analyses considered in this book is to build a data frame with the same structure of the `Tbox`, `Obox`, `Pbox`, `Tadd` and `PbP` datasets (same variables with the same names and same characteristics), introduced in Chapter 2, Section 2.1.

To illustrate how datasets can be manipulated for working with `BasketballAnalyzeR`, we consider the NBA Enhanced Box Score and Standings (2012 - 2018) datasets available on the Kaggle online community of data scientists (https://www.kaggle.com/pablote/nba-enhanced-stats). Many excellent packages for data manipulation are available in R. Here we develop our example using `dplyr`, a popular package which is part of the so-called *tidyverse*, a collection of R packages designed for data science that share a common underlying design philosophy, grammar, and data structures (Wickham and Grolemund, 2016). With `dplyr`, it is possible to perform efficiently complex data manipulation writing elegant and straightforward code. This package is organized around five verbs that cover the majority of the most frequently required data manipulations: *select* certain columns of data, *filter* data to select specific rows, *arrange* the rows of the data matrix into an order, *mutate* data frame to contain new columns, and *summarise* chunks of data in some way. In addition, `dplyr` uses the pipe operator `%>%` of the `magrittr` package. Piping considerably improves the readability of the code and represents a powerful tool for clearly expressing a sequence of multiple operations (Baumer et al., 2017).

The `2012-18_teamBoxScore.csv` file contains box scores data for each of the 82 games played by the 30 NBA teams in the championships from season 2012/2013 to 2017/2018. Each game is recorded in two rows of the dataset, for the home and the away team, respectively. The total number of rows in the dataset is $6 \cdot 30 \cdot 82 - 2 = 14760 - 2 = 14758$ (1 game of the 2012-2013 season is missing). Starting from this dataset, we can generate two datasets that have the same information and structure of `Tbox` and `Obox`. The first step is to read data with `read.csv` and to generate a variable `season` indicating the NBA season.

```
> dts <- read.csv(file="2012-18_teamBoxScore.csv")
> dts$gmDate <- as.Date(as.character(dts$gmDate))
> year <- as.numeric(format(dts$gmDate,"%Y"))
> month <- as.numeric(format(dts$gmDate,"%m"))
> dts$season <- ifelse(month<5, paste0(year-1,"-",year),
                                paste0(year,"-",year+1))
```

Then, for each group defined by season and team (`group_by` function), we calculate (with `summarise`) the 22 variables of the `Tbox` dataset which, for the most part, are simple sums of columns in `2012-18_teamBoxScore.csv`. The code for this data manipulation is simple and intuitive

```
> library(dplyr)
> Tbox2 <- dts %>%
>   group_by(season,  teamAbbr) %>%
>   summarise(GP=n(), MIN=sum(round(teamMin/5)),
      PTS=sum(teamPTS),
      W=sum(teamRslt=="Win"), L=sum(teamRslt=="Loss"),
      P2M=sum(team2PM), P2A=sum(team2PA), P2p=P2M/P2A,
      P3M=sum(team3PM), P3A=sum(team3PA), P3p=P3M/P3A,
      FTM=sum(teamFTM), FTA=sum(teamFTA), FTp=FTM/FTA,
      OREB=sum(teamORB), DREB=sum(teamDRB), AST=sum(teamAST),
      TOV=sum(teamTO), STL=sum(teamSTL), BLK=sum(teamBLK),
      PF=sum(teamPF), PM=sum(teamPTS-opptPTS)) %>%
>   rename(Season=season, Team=teamAbbr) %>%
>   as.data.frame()
```

Similarly, we can generate `Obox` using

```
> Obox2 <- dts %>%
>   group_by(season,  teamAbbr) %>%
>   summarise(GP=n(), MIN=sum(round(opptMin/5)),
      PTS=sum(opptPTS),
      W=sum(opptRslt=="Win"), L=sum(opptRslt=="Loss"),
      P2M=sum(oppt2PM), P2A=sum(oppt2PA), P2p=100*P2M/P2A,
      P3M=sum(oppt3PM), P3A=sum(oppt3PA), P3p=100*P3M/P3A,
      FTM=sum(opptFTM), FTA=sum(opptFTA), FTp=100*FTM/FTA,
      OREB=sum(opptORB), DREB=sum(opptDRB), AST=sum(opptAST),
      TOV=sum(opptTO), STL=sum(opptSTL), BLK=sum(opptBLK),
      PF=sum(opptPF), PM=sum(teamPTS-opptPTS)) %>%
```

```
> 	rename(Season=season, Team=teamAbbr) %>%
> 	as.data.frame()
```

The construction of a dataset similar to `Pbox` requires a different dataset. Game data for each player are available in `2012-18_playerBoxScore.csv`. Using this dataset, and with minor modifications to the code developed for `Tbox` and `Obox`, we can generate the required data frame

```
> dts <- read.csv(file="2012-18_playerBoxScore.csv",
                  encoding="UTF-8")
> dts$gmDate <- as.Date(as.character(dts$gmDate))
> year <- as.numeric(format(dts$gmDate,"%Y"))
> month <- as.numeric(format(dts$gmDate,"%m"))
> dts$season <- ifelse(month<5, paste0(year-1,"-",year),
                       paste0(year,"-",year+1))
> Pbox2 <- dts %>%
>   group_by(season, teamAbbr, playDispNm) %>%
>   summarise(GP=n(), MIN=sum(playMin), PTS=sum(playPTS),
        P2M=sum(play2PM), P2A=sum(play2PA), P2p=100*P2M/P2A,
        P3M=sum(play3PM), P3A=sum(play3PA), P3p=100*P3M/P3A,
        FTM=sum(playFTM), FTA=sum(playFTA), FTp=100*FTM/FTA,
        OREB=sum(playORB), DREB=sum(playDRB), AST=sum(playAST),
        TOV=sum(playTO), STL=sum(playSTL), BLK=sum(playBLK),
        PF=sum(playPF)) %>%
>   rename(Season=season, Team=teamAbbr,
          Player=playDispNm) %>%
>   as.data.frame()
```

It is worth noting that here the groups defined with **group_by** are the single players, differentiated by team and season. In addition, we do not have information about the plus-minus statistics, so the corresponding variable cannot be created in the data frame. On the other hand, we have some additional information such as the players' role and height/weight, which may be added to the **Pbox** data frame and used for further analyses with the **BasketballAnalyzeR** functions.

6.3 CUSTOMIZING PLOTS

The plots created by **BasketballAnalyzeR** are objects generated using the **ggplot2** package, an R package which implements, extends and refines the ideas described in the work of Bertin (1983) and in the "Grammar of Graphics" of Wilkinson (2012). The **ggplot2** package represents a valuable alternative to the standard R graphics. An introduction to the ggplot2 grammar is presented in Wickham (2010). From a practical point of view, one important and distinctive feature of this package is the possibility to store a plot in a ggplot object for modification or future recall. In the following, we will examine some examples (the BasketballAnalyzeR library has to be preliminarily loaded). Let's start by considering the following example

```
> Pbox.sel <- subset(Pbox, MIN>=500)
> attach(Pbox.sel)
> X <- data.frame(AST, TOV, PTSpm=PTS)/MIN
> detach(Pbox.sel)
> mypal <- colorRampPalette(c("blue","yellow","red"))
> p1 <- scatterplot(X, data.var=c("AST","TOV"),
                    z.var="PTSpm", palette=mypal)
> print(p1)
> class(p1)
```

Object p1 is of class ggplot and can be manipulated in different ways to get a customized plot. First, we can add a title plot and change axes labels (using labs), change x-axis range and set x-axis ticks (scale_x_continuous with limits and breaks options, respectively), set panel background color and panel border color (panel.background inside theme), and remove color legend (using guides)

```
> p2 <- p1 +
        labs(title="Scatter plot", x="Assists",
            y="Turnovers") +
        scale_x_continuous(breaks=seq(0,0.35,0.05),
                          limits=c(0,0.35)) +
        theme(panel.background=element_rect(fill="#FFCCCC20",
            colour="red", size=3)) +
        guides(color=FALSE)
> print(p2)
```

It is possible to annotate a plot created with ggplot by adding straight lines, arrows, rectangles and text. With the code reported below, we add an arrow (geom_segment with the arrow option) and text (annotate) at a specific position in the p1 plot

```
> p3 <- p1 +
        geom_segment(x=0.225, y=0.025, xend=X$AST[143]+0.005,
```

```
                yend=X$TOV[143]-0.001, size=1,
                color="red",
                arrow=arrow(length=unit(0.25, "cm"),
                type="closed", angle=20)) +
        annotate("text", x=0.225, y=0.025,
                label=Pbox.sel[143,"Player"],
                color="red", fontface=2, hjust=0)
> print(p3)
```

To add a colored background rectangle to the plot, one can use `geom_rect`

```
> p3 + geom_rect(xmin=0.2, ymin=0.075,
                xmax=Inf, ymax=Inf,
                fill="#DDDDDDAA", color=NA)
```

Using this syntax, the gray rectangle hides the existing points, even if we use a color with transparency. This is due to the fact that the rectangle is at the top layer of the plot. We need to first draw the rectangle and then the points of the scatterplots. This solution can be implemented by placing the gray rectangle at the bottom layer, as follows

```
> p3$layers <- c(geom_rect(xmin=0.2, ymin=0.075,
                        xmax=Inf, ymax=Inf,
                        fill="#DDDDDDAA", color=NA),
                p3$layers)
> print(p3)
```

We can also easily add an image to our plot. In the example that follows, based on **grid** and **jpeg** packages (R Development Core Team, 2008; Urbanek, 2014), we download the NBA logo (in jpeg format) and add it in the left upper area of the original plot using `annotation_custom`

```
> library(grid)
> library(jpeg)
> URL <- "https://goo.gl/WGk6J1"
> download.file(URL,"NBAlogo.jpg", mode="wb")
> NBAlogo <- readJPEG("NBAlogo.jpg", native=TRUE)
> grb <- rasterGrob(NBAlogo, interpolate=TRUE)
> p4 <- p2 +
        annotation_custom(grb, xmin=0.025, xmax=0.05,
```

```
                        ymin=0.1, ymax=0.15) +
         guides(color=FALSE)
> print(p4)
```

Note that https://goo.gl/WGk6J1 is the shortened URL to the link https://cdn.nba.net/nba-drupal-prod/styles/landscape/s3/2017-07/NBA%20Primary%20Logo.jpg.

Another interesting customization available with `ggplot` objects is the possibility to arrange two or more plots on the same page. For example, it is possible to embed one chart within another chart (using the `cowplot` library, Wilke, 2019)

```
> library(cowplot)
> ggdraw() +
    draw_plot(p1) +
    draw_plot(p2, x=0.55, y=0.06, width=0.3, height=0.325)
```

where 0.55 and 0.06 are the x and y locations of the lower left corner of the p3 plot; 0.3 and 0.325 are the width and height of the p3 plot. Another way to plot multiple charts on the same page is to organize the plots on a grid using `grid.arrange` (in the `gridExtra` package) or `plot_grid` (in the `cowplot` package). Throughout the present book, multiple plots were combined together using `grid.arrange`. Below we consider an example that shows how `plot_grid` works

```
> plot_grid(p1, NULL, p2, p4, nrow=2,
          labels = c("A","","B","C"))
```

Here the three plots are arranged on a 2-by-2 grid; the plot placed on the first row and second column of the grid is set to NULL because we want an empty space in this area. The three plots can also be combined in a more complex way; p1 occupies the whole first row, p2 and p3 the second row. In this case, we can specify locations and different sizes using `draw_plot`, as follows

```
> ggdraw() +
    draw_plot(p3, x=0, y=0.5, width=1, height=0.5) +
    draw_plot(p2, x=0, y=0, width=0.5, height=0.5) +
    draw_plot(p4, x=0.5, y=0, width=0.5, height=0.5)
```

Some elements of `ggplot` objects are more difficult to modify, and their customization requires a deeper knowledge about how `ggplot2` works and about the structure of its objects. In the example below, we show how one can change the point shape and size in the p1 scatter plot

```
> q1 <- ggplot_build(p1)
> q1$data[[1]]$shape <- 17
> q1$data[[1]]$size <- 3
> p1b <- ggplot_gtable(q1)
> plot(p1b)
```

The `ggplot_build` function outputs a list of data frames (one for each layer of the `ggplot` object), and a panel object (which contain all information about axis limits, breaks, etc.). In the `p1` scatter plot, there is only one layer and the corresponding data frame in `q1` contains information on x and y coordinates of points, their shape, size, border and fill colors, transparency, etc.

```
> str(q1$data[[1]])
`data.frame':   361 obs. of  10 variables:
 $ colour: chr  "#E1E11D" "#FFAA00" "#DFDF1F" "#C7C737"...
 $ x     : num  0.0869 0.2007 0.1274 0.0549 0.0584 ...
 $ y     : num  0.0775 0.0881 0.0878 0.0588 0.0558 ...
 $ PANEL : Factor w/ 1 level "1": 1 1 1 1 1 1 1 1 1 1 ...
 $ group : int  -1 -1 -1 -1 -1 -1 -1 -1 -1 -1 ...
 $ shape : num  17 17 17 17 17 17 17 17 17 17 ...
 $ size  : num  3 3 3 3 3 3 3 3 3 3 ...
 $ fill  : logi  NA NA NA NA NA NA ...
 $ alpha : logi  NA NA NA NA NA NA ...
 $ stroke: num  0.5 0.5 0.5 0.5 0.5 0.5 0.5 0.5 0.5 0.5...
```

The `ggplot_gtable` function rebuilds all grobs (graphical objects) necessary for displaying the plot.

An interesting package for interactively editing `ggplot2` objects is `ggedit` which, in the author's words, *"is a package that helps users bridge the gap between making a plot and getting all of those pesky plot aesthetics just right, all while keeping everything portable for further research and collaboration"* (Sidi, 2018). `ggedit` can be run from an R console or as a reactive object in any `shiny` application (Chang et al., 2018).

6.4   BUILDING INTERACTIVE GRAPHICS

The R base graphics engine and the `ggplot2` package generate static visualizations of data. A fixed image is required when publishing to a static medium like paper, and can be appropriate when alternate views are not needed. Representing multidimensional data with static images is difficult and often requires to describe a variety of perspectives on the same chart. A significant improvement in computer technology was enabling users to interact with computer displays. A statistical graphic that is either dynamic which means it is capable of smooth motion, or interactive which means it is capable of reaction to the user's action, provides a more natural and more effective way to understand data than that provided by an ordinary static and noninteractive graphic (Young et al., 2011).

`plotly` is a powerful R graphing library for building interactive and dynamic visualizations (Sievert, 2018; Sievert et al., 2017). There are two main ways to build a `plotly` object in R. The `plot_ly()` function transforms data into a `plotly object`, while the `ggplotly()` function transforms a `ggplot` object into a `plotly` object. When a `plotly` object is created, printing it results in an interactive web-based visualization with tooltips, zooming, and panning enabled by default. The R package also has special semantics for arranging, linking, and animating `plotly` objects.

The code below shows how to generate an interactive scatter plot with colored points (the `BasketballAnalyzeR` library has to be preliminarily loaded)

```
> library(plotly)
> Pbox.sel <- subset(Pbox, MIN>=500)
> attach(Pbox.sel)
> X <- data.frame(AST,TOV, PTSpm=PTS)/MIN
> detach(Pbox.sel)
> mypal <- colorRampPalette(c("blue","yellow","red"))
> p5 <- scatterplot(X, data.var=c("AST","TOV"),
                    z.var="PTSpm", palette=mypal)
> ggplotly(p5, tooltip="text")
```

The `ggplotly` command also works for more complex plots, like a matrix of scatter plots

```
> data <- Pbox[1:64, c("PTS","P3M","P2M","OREB","Team")]
> p6 <- scatterplot(data, data.var=1:4, z.var="Team")
> ggplotly(p6)
```

It is worth mentioning that the conversion performed by `ggplotly` from a `ggplot` static plot to a `plotly` interactive graphic is often not perfect and some things might not look exactly the way `ggplot2` does. In addition, sometimes the default interactive properties (e.g., tooltips) might not work the way one wants them to or might take a while to render. Some of these issues can be fixed with a bit of knowledge about `plotly` and the underlying `plotly.js` library. A good tutorial about how to improve `ggplotly` conversions is available here: https://moderndata.plot.ly/learning-from-and-improving-upon-ggplotly-conversions/.

6.5   OTHER R RESOURCES

Several R packages and codes for basketball have been developed in the last years.

Some of these projects are devoted to scraping basketball data from sports websites. For example:

- `ballr` (by Ryan Elmore and Peter DeWitt), an R package that provides simple functions for accessing data and tables available on http://www.basketball-reference.com
  `https://CRAN.R-project.org/package=ballr`

- `bbr` (by Max Joseph), an R package to scrape data from basketball-reference.com
  `https://github.com/mbjoseph/bbr`

- `nbaTools` (by Chirag Agrawal), an R package for scraping NBA related data from NBA.com
  `https://github.com/ccagrawal/nbaTools`

- `ncaahoopR` (by Luke Benz), an R package for working with NCAA Basketball Play-by-Play Data
  `https://github.com/lbenz730/ncaahoopR`

Other packages offer R wrapper functions for downloading data from commercial websites using application program interface (API):

- `NBAloveR` (by Koki Ando), an interface to the online basketball data resources such as Basketball reference API https://www.basketball-reference.com/ and helps R users analyze basketball data
  `https://cran.r-project.org/web/packages/NBAloveR/index.html`

- `NBAr` (by Patrick Chodowski), a set of wrapper functions for downloading and simple processing of data from http://stats.nba.com API
  `https://github.com/PatrickChodowski/NBAr`

- `mysportsfeedsR` (by @MySportsFeeds), an R wrapper functions for the MySportsFeeds Sports Data API
  https://www.mysportsfeeds.com
  `https://github.com/MySportsFeeds/mysportsfeeds-r`

- `stattleshipR` (by @stattleship), Stattleship R Wrapper
  https://api.stattleship.com/
  `https://github.com/stattleship/stattleship-r`

In addition, a few packages are designed to provide tools for basketball data analysis and visualization. Two interesting examples are:

- `BallR` (by Todd W. Schneider), Interactive NBA and NCAA Shot Charts with R and Shiny
  `https://github.com/toddwschneider/ballr`

- `NBA_SportVu` (by Rajiv Shah), R code for exploring the NBA SportVu motion data
  `https://github.com/rajshah4/NBA_SportVu`

Many of these resources are available on `github.com`. A partial list can be retrieved at this link:

`https://github.com/search?q=%22basketball%22+%5BR%5D&type=Repositories`

Bibliography

Aglioti, S. M., Cesari, P., Romani, M., and Urgesi, C. (2008). Action anticipation and motor resonance in elite basketball players. *Nature Neuroscience*, 11(9):1109.

Alagappan, M. (2012). From 5 to 13: Redefining the positions in basketball. In *2012 MIT Sloan Sports Analytics Conference*. *http://www.sloansportsconference.com*.

Alamar, B. C. (2013). *Sports analytics: A guide for coaches, managers, and other decision makers*. Columbia University Press.

Albert, J., Glickman, M. E., Swartz, T. B., and Koning, R. H. (2017). *Handbook of Statistical Methods and Analyses in Sports*. CRC Press.

Allison, P. D. (2001). *Missing data*, volume 136. Sage publications.

Annis, D. H. (2006). Optimal end-game strategy in basketball. *Journal of Quantitative Analysis in Sports*, 2(2):1.

Ante, P., Slavko, T., and Igor, J. (2014). Interdependencies between defence and offence in basketball. *Sport Science*, 7(2):62–66.

Araújo, D. and Davids, K. (2016). Team synergies in sport: theory and measures. *Frontiers in Psychology*, 7.

Araújo, D., Davids, K., and Hristovski, R. (2006). The ecological dynamics of decision making in sport. *Psychology of Sport and Exercise*, 7(6):653–676.

Araújo, D., Davids, K. W., Chow, J. Y., Passos, P., and Raab, M. (2009). The development of decision making skill in sport: an ecological dynamics perspective. In *Perspectives on cognition and action in sport*, pages 157–169. Nova Science Publishers, Inc.

Araújo, D. and Esteves, P. (2010). The irreducible variability of decision making in basketball. *Aportaciones teóricas y prácticas para el baloncesto del futuro*, page 171.

Arendt, E. and Dick, R. (1995). Knee injury patterns among men and women in collegiate basketball and soccer: NCAA data and review of literature. *The American Journal of Sports Medicine*, 23(6):694–701.

Arkes, J. (2010). Revisiting the hot hand theory with free throw data in a multivariate framework. *Journal of Quantitative Analysis in Sports*, 6(1):1–12.

Ashtiani, M. (2019). CINNA*: Deciphering Central Informative Nodes in Network Analysis*. R package version 1.1.53.

Auguie, B. (2017). *gridExtra: Miscellaneous Functions for "Grid" Graphics*. R package version 2.3.

Avugos, S., Köppen, J., Czienskowski, U., Raab, M., and Bar-Eli, M. (2013). The "hot hand" reconsidered: A meta-analytic approach. *Psychology of Sport and Exercise*, 14(1):21–27.

Bache, S. M. and Wickham, H. (2014). `magrittr`*: A Forward-Pipe Operator for* R. R package version 1.5.

Bar-Eli, M., Avugos, S., and Raab, M. (2006). Twenty years of "hot hand" research: Review and critique. *Psychology of Sport and Exercise*, 7(6):525–553.

Baumer, B. S., Kaplan, D. T., and Horton, N. J. (2017). *Modern data science with R*. CRC Press.

Bertin, J. (1983). Semiology of graphics (English translation by William J Berg). *Madison, WI: The University of Wisconsin Press. (Original work published 1967)*.

Bianchi, F. (2016). *Towards a new meaning of modern basketball players positions*. Master Degree Thesis in Computer Science and Multimedia, University of Pavia, Italy.

Bianchi, F., Facchinetti, T., and Zuccolotto, P. (2017). Role revolution: towards a new meaning of positions in basketball. *Electronic Journal of Applied Statistical Analysis*, 10(3):712–734.

Borg, I., Groenen, P. J., and Mair, P. (2017). *Applied multidimensional scaling and unfolding*. Springer.

Bornn, L., Cervone, D., Franks, A., and Miller, A. (2017). Studying basketball through the lens of player tracking data. In *Handbook of Statistical Methods and Analyses in Sports*, pages 245–269. Chapman & Hall/CRC.

Bourbousson, J., Poizat, G., Saury, J., and Sève, C. (2010a). Team coordination in basketball: Description of the cognitive connections among teammates. *Journal of Applied Sport Psychology*, 22(2):150–166.

Bourbousson, J., Sève, C., and McGarry, T. (2010b). Space–time coordination dynamics in basketball: Part 1. intra-and inter-couplings among player dyads. *Journal of Sports Sciences*, 28(3):339–347.

Bourbousson, J., Sève, C., and McGarry, T. (2010c). Space–time coordination dynamics in basketball: Part 2. the interaction between the two teams. *Journal of Sports Sciences*, 28(3):349–358.

Breiman, L. (1996). Bagging predictors. *Machine Learning*, 24(2):123–140.

Breiman, L. (2001a). Random forests. *Machine Learning*, 45(1):5–32.

Breiman, L. (2001b). Statistical modeling: The two cultures (with comments and a rejoinder by the author). *Statistical Science*, 16(3):199–231.

Breiman, L., Friedman, J. H., Stone, C. J., and Olshen, R. A. (1984). *Classification and regression trees*. CRC press.

Brown, M. and Sokol, J. (2010). An improved LRMC method for NCAA basketball prediction. *Journal of Quantitative Analysis in Sports*, 6(3):1–23.

Butts, C. T. (2008). `network`: a package for managing relational data in R. *Journal of Statistical Software*, 24(2).

Butts, C. T. (2015). `network`: *Classes for Relational Data*. The Statnet Project. R package version 1.13.0.1.

Butts, D. (1986). NCAA adds three-point basket. *The Bryan Times*.

Carlsson, G. (2009). Topology and data. *Bulletin of the American Mathematical Society*, 46(2):255–308.

Cervone, D., D'Amour, A., Bornn, L., and Goldsberry, K. (2016). A multiresolution stochastic process model for predicting basketball possession outcomes. *Journal of the American Statistical Association*, 111(514):585–599.

Chambers, J. (2008). *Software for data analysis: programming with R*. Springer Science & Business Media.

Chang, W., Cheng, J., Allaire, J., Xie, Y., and McPherson, J. (2018). *shiny: Web Application Framework for R*. R package version 1.2.0.

Chen, X., Ye, Y., Williams, G., and Xu, X. (2007). A survey of open source data mining systems. In *Pacific-Asia Conference on Knowledge Discovery and Data Mining*, pages 3–14. Springer.

Clemente, F. M., Martins, F. M. L., Kalamaras, D., and Mendes, R. S. (2015). Network analysis in basketball: inspecting the prominent players using centrality metrics. *Journal of Physical Education and Sport*, 15(2):212.

Cleveland, W. S. (1979). Robust locally weighted regression and smoothing scatterplots. *Journal of the American Statistical Association*, 74(368):829–836.

Cleveland, W. S. and Devlin, S. J. (1988). Locally weighted regression: an approach to regression analysis by local fitting. *Journal of the American Statistical Association*, 83(403):596–610.

Cobb, G. W. and Moore, D. S. (1997). Mathematics, statistics, and teaching. *The American Mathematical Monthly*, 104(9):801–823.

Cooper, W. W., Ruiz, J. L., and Sirvent, I. (2009). Selecting non-zero weights to evaluate effectiveness of basketball players with DEA. *European Journal of Operational Research*, 195(2):563–574.

Corsello, A. (2016). The stunning, strange, beautiful game of Manuel Neuer. http://www.espn.com/espn/feature/story/_/page/enterprise-neuer160525/bayern-munich-manuel-neuer-changing-means-goalie.

Courneya, K. S. and Carron, A. V. (1992). The home advantage in sport competitions: A literature review. *Journal of Sport and Exercise Psychology*, 14(1):13–27.

Cox, T. F. and Cox, M. A. (2000). *Multidimensional scaling*. Chapman & hall/CRC.

Csardi, G. and Nepusz, T. (2006). The `igraph` software package for complex network research. *InterJournal*, Complex Systems:1695.

Csataljay, G., O'Donoghue, P., Hughes, M., and Dancs, H. (2009). Performance indicators that distinguish winning and losing teams in basketball. *International Journal of Performance Analysis in Sport*, 9(1):60–66.

De Leeuw, J. and Mair, P. (2009). Gifi methods for optimal scaling in R: The package homals. *Journal of Statistical Software*, pages 1–30.

de Oliveira, R. F., Oudejans, R. R., and Beek, P. J. (2006). Late information pick-up is preferred in basketball jump shooting. *Journal of Sport Sciences*, 24:933–940.

De Rose, D. J. (2004). Statistical analysis of basketball performance indicators according to home/away games and winning and losing teams. *Journal of Human Movement Studies*, 47:327–336.

Deshpande, S. K. and Jensen, S. T. (2016). Estimating an NBA player's impact on his team's chances of winning. *Journal of Quantitative Analysis in Sports*, 12(2):51–72.

Dirks, K. T. (2000). Trust in leadership and team performance: Evidence from NCAA basketball. *Journal of Applied Psychology*, 85(6):1004.

Engelmann, J. (2017). Possession-based player performance analysis in basketball (adjusted +/− and related concepts). In *Handbook of Statistical Methods and Analyses in Sports*, pages 215–227. Chapman & Hall/CRC.

Erčulj, F. and Štrumbelj, E. (2015). Basketball shot types and shot success in different levels of competitive basketball. *PloS one*, 10(6):e0128885.

Fagerland, M., Lydersen, S., and Laake, P. (2017). *Statistical Analysis of Contingency Tables.* CRC Press.

Fahrmeir, L. and Tutz, G. (2013). *Multivariate statistical modelling based on generalized linear models.* Springer Science & Business Media.

Fearnhead, P. and Taylor, B. M. (2011). On estimating the ability of NBA players. *Journal of Quantitative Analysis in Sports*, 7(3).

Fewell, J. H., Armbruster, D., Ingraham, J., Petersen, A., and Waters, J. S. (2012). Basketball teams as strategic networks. *PloS one*, 7(11):e47445.

Flannery, P. (2016). Draymond Green is redefining NBA stardom. Even he didn't see that coming. http://www.sbnation.com/2016/2/16/10987022/draymond-green-warriors-nba-unexpected-star.

Forgy, E. (1965). Cluster analysis of multivariate data: Efficiency vs. interpretability of classification. *Biometrics*, 21(3):768–769.

Franks, A. M., D'Amour, A., Cervone, D., and Bornn, L. (2016). Meta-analytics: tools for understanding the statistical properties of sports metrics. *Journal of Quantitative Analysis in Sports*, 12(4):151–165.

Friedman, J. H. (2001). Greedy function approximation: a gradient boosting machine. *Annals of Statistics*, 29(5):1189–1232.

Friedman, J. H., Hastie, T., and Tibshirani, R. (2009). *The elements of statistical learning.* Springer series in statistics, New York, NY, USA.

Friedman, J. H. and Popescu, B. E. (2008). Predictive learning via rule ensembles. *The Annals of Applied Statistics*, 2(3):916–954.

Fujimura, A. and Sugihara, K. (2005). Geometric analysis and quantitative evaluation of sport teamwork. *Systems and Computers in Japan*, 36(6):49–58.

Gabel, A. and Redner, S. (2012). Random walk picture of basketball scoring. *Journal of Quantitative Analysis in Sports*, 8(1):1416.

García, J., Ibáñez, S. J., De Santos, R. M., Leite, N., and Sampaio, J. (2013). Identifying basketball performance indicators in regular season and playoff games. *Journal of Human Kinetics*, 36(1):161–168.

Gifi, A. (1990). *Nonlinear multivariate analysis*. John Wiley & Sons Ltd.

Gilovich, T., Vallone, R., and Tversky, A. (1985). The hot hand in basketball: On the misperception of random sequences. *Cognitive Psychology*, 17(3):295–314.

Goldman, M. and Rao, J. M. (2012). Effort vs. concentration: the asymmetric impact of pressure on NBA performance. In MIT *Sloan Sports Analytics Conference*.

Golfarelli, M. and Rizzi, S. (2009). *Data warehouse design: Modern principles and methodologies*, volume 5. McGraw-Hill New York.

Gómez Sánchez, J., Moll Sotomayor, J. A., and Pila Teleña, A. (1980). Baloncesto: Técnica de entrenamiento y dirección de equipo. *Madrid, Pila Teleña*.

Greenacre, M. (2017). *Correspondence Analysis in Practice*. Chapman & Hall/CRC, New York, 3rd edition.

Gudmundsson, J. and Horton, M. (2016). Spatio-temporal analysis of team sports. a survey. *arXiv preprint arXiv:1602.06994*.

Gupta, A. A. (2015). A new approach to bracket prediction in the NCAA men's basketball tournament based on a dual-proportion likelihood. *Journal of Quantitative Analysis in Sports*, 11(1):53–67.

Guyon, I., Von Luxburg, U., and Williamson, R. C. (2009). Clustering: Science or art. In *NIPS 2009 workshop on clustering theory*, pages 1–11.

Hamilton, J. D. (2010). Regime switching models. In *Macroeconometrics and Time Series Analysis*, pages 202–209. Springer.

Han, J., Pei, J., and Kamber, M. (2011). *Data mining: concepts and techniques*. Elsevier.

Hand, D. J. (2008). *Statistics: a very short introduction*, volume 196. Oxford University Press.

Hand, D. J. and Taylor, C. C. (1987). *Multivariate Analysis of Variance and Repeated Measures: A Practical Approach for Behavioural Scientists*, volume 5. Chapman & Hall/CRC Texts in Statistical Science.

Härdle, W. K. (1990). *Applied nonparametric regression*. Cambridge University Press.

Härdle, W. K. and Simar, L. (2015). *Applied multivariate statistical analysis*. Springer-Verlag.

Hartigan, J. A. and Wong, M. A. (1979). Algorithm as 136: A k-means clustering algorithm. *Journal of the Royal Statistical Society. Series C (Applied Statistics)*, 28(1):100–108.

Hawkins, D. M. (1980). *Identification of outliers*, volume 11. Springer.

Hawkins, D. M. (2004). The problem of overfitting. *Journal of chemical information and computer sciences*, 44(1):1–12.

Hennig, C., Meila, M., Murtagh, F., and Rocci, R. (2015). *Handbook of Cluster Analysis*. CRC Press.

Hernández, M. A. and Stolfo, S. J. (1998). Real-world data is dirty: Data cleansing and the merge/purge problem. *Data Mining and Knowledge Discovery*, 2(1):9–37.

Heuzé, J.-P., Raimbault, N., and Fontayne, P. (2006). Relationships between cohesion, collective efficacy and performance in professional basketball teams: An examination of mediating effects. *Journal of Sports Sciences*, 24(1):59–68.

Hollander, M. and Wolfe, D. A. (1999). *Nonparametric statistical methods*. Wiley-Interscience.

Hosmer, D. W. J., Lemeshow, S., and Sturdivant, R. X. (2013). *Applied Logistic Regression*. John Wiley & Sons.

Ibáñez, S. J., García, J., Feu, S., Lorenzo, A., and Sampaio, J. (2009). Effects of consecutive basketball games on the game-related statistics that discriminate winner and losing teams. *Journal of Sports Science and Medicine*, 8(3):458–462.

Ibáñez, S. J., Sampaio, J., Sáenz-López Buñuel, P., Giménez Fuentes-Guerra, J., and Janeira, M. (2003). Game statistics discriminating the final outcome of junior world basketball championship matches (Portugal 1999). *Journal of Human Movement Studies*, 45(1):1–20.

Ihaka, R. and Gentleman, R. (1996). R: a language for data analysis and graphics. *Journal of Computational and Graphical Statistics*, 5(3):299–314.

Johnson, R. A. and Wichern, D. W. (2013). *Applied Multivariate Statistical Analysis*. Pearson Education Limited, New York.

Jolliffe, I. (1986). *Principal Component Analysis*. Springer-Verlag.

Kamada, T. and Kawai, S. (1989). An algorithm for drawing general undirected graphs. *Information Processing Letters*, 31(1):7–15.

Kaufman, L. and Rousseeuw, P. J. (1990). *Finding groups in data: an introduction to cluster analysis*. John Wiley & Sons, New York.

Kay, H. K. (1966). *A statistical analysis of the profile technique for the evaluation of competitive basketball performance*. PhD thesis, University of Alberta.

Kenett, R. S. and Redman, T. C. (2019). *The Real Work of Data Science: Turning data into information, better decisions, and stronger organizations*. John Wiley & Sons.

Kenett, R. S. and Shmueli, G. (2016). *Information quality (InfoQ): The Potential of Data and Analytics to Generate Knowledge*. John Wiley & Sons.

Kim, W., Choi, B.-J., Hong, E.-K., Kim, S.-K., and Lee, D. (2003). A taxonomy of dirty data. *Data Mining and Knowledge Discovery*, 7(1):81–99.

Koehler, J. J. and Conley, C. A. (2003). The "hot hand" myth in professional basketball. *Journal of Sport and Exercise Psychology*, 25(2):253–259.

Koh, K. T., Wang, C. K. J., and Mallett, C. (2011). Discriminating factors between successful and unsuccessful teams: A case study in elite youth Olympic basketball games. *Journal of Quantitative Analysis in Sports*, 7(3).

Koh, K. T., Wang, C. K. J., and Mallett, C. (2012). Discriminating factors between successful and unsuccessful elite youth Olympic female basketball teams. *International Journal of Performance Analysis in Sport*, 12(1):119–131.

Kohonen, T. (1982). Self-organized formation of topologically correct feature maps. *Biological Cybernetics*, 43(1):59–69.

Kohonen, T. (1990). The self-organizing map. *Proceedings of the IEEE*, 78(9):1464–1480.

Kruskal, J. B. (1964a). Multidimensional scaling by optimizing goodness of fit to a nonmetric hypothesis. *Psychometrika*, 29(1):1–27.

Kruskal, J. B. (1964b). Nonmetric multidimensional scaling: a numerical method. *Psychometrika*, 29(2):115–129.

Kruskal, J. B. and Wish, M. (1978). Multidimensional scaling. *Sage University Paper Series on Quantitative Applications in the Social Sciences*, 7(11).

Kubatko, J., Oliver, D., Pelton, K., and Rosenbaum, D. T. (2007). A starting point for analyzing basketball statistics. *Journal of Quantitative Analysis in Sports*, 3(3):1–22.

Lamas, L., De Rose Jr., D., Santana, F. L., Rostaiser, E., Negretti, L., and Ugrinowitsch, C. (2011). Space creation dynamics in basketball offence: validation and evaluation of elite teams. *International Journal of Performance Analysis in Sport*, 11(1):71–84.

Larose, D. T. and Larose, C. D. (2014). *Discovering knowledge in data: an introduction to data mining*. John Wiley & Sons.

Liebetrau, A. M. (1983). *Measures of association*. Sage, Newbury Park, CA.

Little, R. J. and Rubin, D. B. (2014). *Statistical analysis with missing data*, volume 333. John Wiley & Sons.

Lloyd, S. (1982). Least squares quantization in pcm. *IEEE Transactions on Information Theory*, 28(2):129–137.

Loeffelholz, B., Bednar, E., and Bauer, K. W. (2009). Predicting NBA games using neural networks. *Journal of Quantitative Analysis in Sports*, 5(1):1–15.

Lopez, M. J. and Matthews, G. J. (2015). Building an NCAA men's basketball predictive model and quantifying its success. *Journal of Quantitative Analysis in Sports*, 11(1):5–12.

Lorenzo, A., Gómez, M. Á., Ortega, E., Ibáñez, S. J., and Sampaio, J. (2010). Game related statistics which discriminate between winning and losing under-16 male basketball games. *Journal of Sports Science & Medicine*, 9(4):664.

Lynch, J. (1987). High school basketball draws line, adopts 3-point rule. *Los Angeles Times*. Friday, March 27.

MacQueen, J. (1967). Some methods for classification and analysis of multivariate observations. In *Proceedings of the fifth Berkeley symposium on mathematical statistics and probability*, volume 1, pages 281–297. Oakland, CA, USA.

Madden, C. C., Kirkby, R. J., McDonald, D., Summers, J. J., Brown, D. F., and King, N. J. (1995). Stressful situations in competitive basketball. *Australian Psychologist*, 30(2):119–124.

Madden, C. C., Summers, J. J., and Brown, D. F. (1990). The influence of perceived stress on coping with competitive basketball. *International Journal of Sport Psychology*, 21(1):21–35.

Maddi, S. R. and Hess, M. J. (1992). Personality hardiness and success in basketball. *International journal of sport psychology*.

Mahmood, Z. (2015). How 'The Makelele Role' redefined English football. http://www.sportskeeda.com/football/how-the-makelele-role-redefined-english-football.

Manner, H. (2016). Modeling and forecasting the outcomes of NBA basketball games. *Journal of Quantitative Analysis in Sports*, 12(1):31–41.

Matejka, J. and Fitzmaurice, G. (2017). Same stats, different graphs: Generating datasets with varied appearance and identical statistics through simulated annealing. In *Proceedings of the 2017 CHI Conference on Human Factors in Computing Systems*, CHI '17, pages 1290–1294, New York, NY, USA. ACM.

Matloff, N. (2011). *The art of R programming: A tour of statistical software design*. No Starch Press.

McCullagh, P. (2002). What is a statistical model? *Annals of Statistics*, 30(5):1225–1267.

McCullagh, P. and Nelder, J. A. (1989). *Generalized linear models*, volume 37. CRC Press.

Merlo, J. and Lynch, K. (2010). Association, measures of. In *Encyclopedia of Research Design*, pages 47–51. SAGE Publications, Inc.

Metulini, R., Manisera, M., and Zuccolotto, P. (2017a). Sensor analytics in basketball. In *Proceedings of the 6th International Conference on Mathematics in Sport*.

Metulini, R., Manisera, M., and Zuccolotto, P. (2017b). Space-time analysis of movements in basketball using sensor data. In *Statistics and Data Science: new challenges, new generations - SIS2017 proceeding*.

Metulini, R., Manisera, M., and Zuccolotto, P. (2018). Modelling the dynamic pattern of surface area in basketball and its effects on team performance. *Journal of Quantitative Analysis in Sports*, 14(3):117–130.

Meyer, D., Zeileis, A., and Hornik, K. (2017). vcd*: Visualizing Categorical Data*. R package version 1.4-4.

Meyers, A. W. and Schleser, R. (1980). A cognitive behavioral intervention for improving basketball performance. *Journal of Sport Psychology*, 2(1):69–73.

Miller, A. C. and Bornn, L. (2017). Possession sketches: Mapping NBA strategies. *MIT Sloan Sports Analytics Conference 2017*.

Miller, S. and Bartlett, R. (1996). The relationship between basketball shooting kinematics, distance and playing. *Journal of Sports Sciences*, 14(3):243–253.

Miller, T. W. (2015). *Sports analytics and data science: winning the game with methods and models*. FT Press.

Milligan, G. W. (1996). Clustering validation: results and implications for applied analysis. In Arabie, P., Hubert, L. J., and De Soete, G., editors, *Clustering and classification*, pages 341–375. World Scientific, Singapore.

Milligan, G. W. and Cooper, M. C. (1985). An examination of procedures for determining the number of clusters in a data set. *Psychometrika*, 50:159–179.

Monson, S. (2012). Mobile Quarterbacks Redefining the Position. http://bleacherreport.com/articles/1251778-mobile-quarterbacks-redefining-the-position.

Murrell, P. (2016). *R graphics*. CRC Press.

Navarro, D. (2015). *Learning statistics with R: A tutorial for psychology students and other beginners. (Version 0.5)*. University of Adelaide, Adelaide, Australia. R package version 0.5.

Newman, M. (2018). *Networks*. Oxford University Press.

Noecker, C. A. and Roback, P. (2012). New insights on the tendency of NCAA basketball officials to even out foul calls. *Journal of Quantitative Analysis in Sports*, 8(3):1–23.

Okubo, H. and Hubbard, M. (2006). Dynamics of the basketball shot with application to the free throw. *Journal of Sports Sciences*, 24(12):1303–1314.

Oliver, D. (2004). *Basketball on paper: rules and tools for performance analysis*. Potomac Books, Inc.

Özmen, U. M. (2012). Foreign player quota, experience and efficiency of basketball players. *Journal of Quantitative Analysis in Sports*, 8(1):1–18.

Page, G. L., Barney, B. J., and McGuire, A. T. (2013). Effect of position, usage rate, and per game minutes played on NBA player production curves. *Journal of Quantitative Analysis in Sports*, 9(4):337–345.

Page, G. L., Fellingham, G. W., and Reese, S. C. (2007). Using boxscores to determine a position's contribution to winning basketball games. *Journal of Quantitative Analysis in Sports*, 3(4):1.

Passos, P., Araújo, D., and Volossovitch, A. (2016). *Performance analysis in team sports*. Taylor & Francis.

Passos, P., Davids, K., Araújo, D., Paz, N., Minguéns, J., and Mendes, J. (2011). Networks as a novel tool for studying team ball sports as complex social systems. *Journal of Science and Medicine in Sport*, 14(2):170–176.

Patel, R. (2017). *Clustering Professional Basketball Players by Performance*. PhD thesis, UCLA.

Pedersen, T. L. (2019). tidygraph*: A Tidy* API *for Graph Manipulation*. R package version 1.1.2.

Perica, A., Trninić, S., and Jelaska, I. (2011). Introduction into the game states analysis system in basketball. *Fizička kultura*, 65(2):51–78.

Perše, M., Kristan, M., Kovačič, S., Vučkovič, G., and Perš, J. (2009). A trajectory-based analysis of coordinated team activity in a basketball game. *Computer Vision and Image Understanding*, 113(5):612–621.

Piette, J., Anand, S., and Zhang, K. (2010). Scoring and shooting abilities of NBA players. *Journal of Quantitative Analysis in Sports*, 6(1).

Piette, J., Pham, L., and Anand, S. (2011). Evaluating basketball player performance via statistical network modeling. In *MIT Sloan Sports Anal. Conf.*

R Development Core Team (2008). *R: A Language and Environment for Statistical Computing*. R Foundation for Statistical Computing, Vienna, Austria. ISBN 3-900051-07-0.

Rahm, E. and Do, H. H. (2000). Data cleaning: Problems and current approaches. *IEEE Data Eng. Bull.*, 23(4):3–13.

Ross, S. D. (2007). Segmenting sport fans using brand associations: A cluster analysis. *Sport Marketing Quarterly*, 16(1):15.

Ruiz, F. J. and Perez-Cruz, F. (2015). A generative model for predicting outcomes in college basketball. *Journal of Quantitative Analysis in Sports*, 11(1):39–52.

Sampaio, J., Drinkwater, E. J., and Leite, N. M. (2010a). Effects of season period, team quality, and playing time on basketball players' game-related statistics. *European Journal of Sport Science*, 10(2):141–149.

Sampaio, J. and Janeira, M. (2003). Statistical analyses of basketball team performance: understanding teams' wins and losses according to a different index of ball possessions. *International Journal of Performance Analysis in Sport*, 3(1):40–49.

Sampaio, J., Janeira, M., Ibáñez, S., and Lorenzo, A. (2006). Discriminant analysis of game-related statistics between basketball guards, forwards and centres in three professional leagues. *European Journal of Sport Science*, 6(3):173–178.

Sampaio, J., Lago, C., and Drinkwater, E. J. (2010b). Explanations for the United States of America's dominance in basketball at the Beijing Olympic Games (2008). *Journal of Sports Sciences*, 28(2):147–152.

Sanders, S. (1981). 22 will get you 3. *Spartanburg Herald*.

Sarle, W. S. (1996). Stopped training and other remedies for overfitting. *Computing science and statistics*, pages 352–360.

Schumaker, R. P., Solieman, O. K., and Chen, H. (2010). *Sports Data Mining*. Springer.

Schwarz, W. (2012). Predicting the maximum lead from final scores in basketball: A diffusion model. *Journal of Quantitative Analysis in Sports*, 8(4).

Seifriz, J. J., Duda, J. L., and Chi, L. (1992). The relationship of perceived motivational climate to intrinsic motivation and beliefs about success in basketball. *Journal of Sport and Exercise Psychology*, 14:375–392.

Severini, T. A. (2014). *Analytic methods in sports: Using mathematics and statistics to understand data from baseball, football, basketball, and other sports*. Chapman & Hall/CRC.

Shortridge, A., Goldsberry, K., and Adams, M. (2014). Creating space to shoot: quantifying spatial relative field goal efficiency in basketball. *Journal of Quantitative Analysis in Sports*, 10(3):303–313.

Sidi, J. (2018). **ggedit***: Interactive 'ggplot2' Layer and Theme Aesthetic Editor. R package version 0.3.0.*

Sievert, C., Parmer, C., Hocking, T., Chamberlain, S., Ram, K., Corvellec, M., and Despouy, P. (2017). plotly: *Create Interactive Web Graphics via 'plotly.js'*. R package version 4.7.1.

Sievert, Carson (2018). plotly *for* R. https://plotly-book.cpsievert.me.

Silverman, B. W. (2018). *Density estimation for statistics and data analysis*. Routledge.

Sims, C. A. (1980). Macroeconomics and reality. *Econometrica: Journal of the Econometric Society*, 48(1):1–48.

Skinner, B. (2010). The price of anarchy in basketball. *Journal of Quantitative Analysis in Sports*, 6(1).

Skinner, B. and Goldman, M. (2017). Optimal strategy in basketball. In *Handbook of Statistical Methods and Analyses in Sports*, pages 229–244. Chapman & Hall/CRC.

Sollich, P. and Krogh, A. (1996). Learning with ensembles: How overfitting can be useful. In *Advances in neural information processing systems*, pages 190–196.

Taylor, J. (1987). Predicting athletic performance with self-confidence and somatic and cognitive anxiety as a function of motor and physiological requirements in six sports. *Journal of Personality*, 55(1):139–153.

Therón, R. and Casares, L. (2010). Visual analysis of time-motion in basketball games. In *Smart Graphics*, pages 196–207. Springer.

Travassos, B., Araújo, D., Davids, K., Esteves, P. T., and Fernandes, O. (2012). Improving passing actions in team sports by developing interpersonal interactions between players. *International Journal of Sports Science & Coaching*, 7(4):677–688.

Trninić, S., Dizdar, D., and Lukšić, E. (2002). Differences between winning and defeated top quality basketball teams in final tournaments of European club championship. *Collegium Antropologicum*, 26(2):521–531.

Tversky, A. and Gilovich, T. (2005). The cold facts about the "hot hand" in basketball. *Anthology of Statistics in Sports*, 16:169.

Urbanek, S. (2014). jpeg: *Read and write* JPEG *images*. R package version 0.1-8.

Vergin, R. (2000). Winning streaks in sports and the misperception of momentum. *Journal of Sport Behavior*, 23(2):181.

Vickers, W. D. (2006). *Multi-level integrated classifications based on the 2001 census*. PhD thesis, University of Leeds.

Vračar, P., Štrumbelj, E., and Kononenko, I. (2016). Modeling basketball play-by-play data. *Expert Systems with Applications*, 44:58–66.

Warner, R. M. (2013). *Applied Statistics: From Bivariate Through Multivariate Techniques*. Sage Publications, Inc.

Weiss, M. R. and Friedrichs, W. D. (1986). The influence of leader behaviors, coach attributes, and institutional variables on performance and satisfaction of collegiate basketball teams. *Journal of Sport Psychology*, 8(4):332–346.

West, B. T. (2008). A simple and flexible rating method for predicting success in the NCAA basketball tournament: Updated results from 2007. *Journal of Quantitative Analysis in Sports*, 4(2):8.

Wickham, H. (2010). A layered grammar of graphics. *Journal of Computational and Graphical Statistics*, 19(1):3–28.

Wickham, H. (2014). *Advanced R*. CRC Press.

Wickham, H. (2016). ggplot2: *Elegant Graphics for Data Analysis*. Springer-Verlag New York.

Wickham, H., François, R., Henry, L., and Müller, K. (2019). dplyr: *A Grammar of Data Manipulation*. R package version 0.8.0.1.

Wickham, H. and Grolemund, G. (2016). *R for data science: import, tidy, transform, visualize, and model data*. O'Reilly Media, Inc.

Wilke, C. O. (2019). cowplot: *Streamlined Plot Theme and Plot Annotations for 'ggplot2'*. R package version 0.9.4.

Wilkinson, L. (2012). The grammar of graphics. In Gentle, J. E., Härdle, W. K., and Mori, Y., editors, *Handbook of Computational Statistics*, pages 375–414. Springer, Berlin, Heidelberg.

Winston, W. L. (2012). *Mathletics: How gamblers, managers, and sports enthusiasts use mathematics in baseball, basketball, and football*. Princeton University Press.

Witten, I. H., Frank, E., Hall, M. A., and Pal, C. J. (2016). *Data Mining: Practical machine learning tools and techniques*. Morgan Kaufmann.

Wu, S. and Bornn, L. (2018). Modeling offensive player movement in professional basketball. *The American Statistician*, 72(1):72–79.

Young, F. W., Valero-Mora, P. M., and Friendly, M. (2011). *Visual statistics: seeing data with dynamic interactive graphics*, volume 914. John Wiley & Sons.

Yuan, L.-H., Liu, A., Yeh, A., Kaufman, A., Reece, A., Bull, P., Franks, A., Wang, S., Illushin, D., and Bornn, L. (2015). A mixture-of-modelers approach to forecasting NCAA tournament outcomes. *Journal of Quantitative Analysis in Sports*, 11(1):13–27.

Zha, T. (2010). Vector autoregressions. In Durlauf, S. N. and Blume, L. E., editors, *Macroeconometrics and Time Series Analysis*, pages 378–390. Springer.

Zhang, T., Hu, G., and Liao, Q. (2013). Analysis of offense tactics of basketball games using link prediction. In *Computer and Information Science (ICIS), 2013 IEEE/ACIS 12th International Conference on*, pages 207–212. IEEE.

Zuccolotto, P., Manisera, M., and Sandri, M. (2018). Big data analytics for modeling scoring probability in basketball: The effect of shooting under high-pressure conditions. *International Journal of Sports Science & Coaching*, 13(4):569–589.

Index